THE MULTIVERSE

THE MULTIVERSE

Uncovering a universal mystery

BRIAN CLEGG

Published in the UK and USA in 2025 by
Icon Books Ltd, Omnibus Business Centre,
39–41 North Road, London N7 9DP
email: info@iconbooks.com
www.iconbooks.com

ISBN: 978-183773-235-7
eBook: 978-183773-236-4

Typeset by SJmagic DESIGN SERVICES, India.

Printed and bound in the UK.

Appointed GPSR EU Representative:
Easy Access System Europe Oü, 16879218
Address: Mustamäe tee 50, 10621, Tallinn, Estonia
Contact Details: gpsr.requests@easproject.com, +358 40 500 3575

For Gillian, Chelsea and Rebecca

ACKNOWLEDGEMENTS

Cosmology has always fascinated and frustrated me in equal measures, since well before doing a Natural Sciences degree specialising in physics. As a huge fan of Fred Hoyle in my teens I was disappointed when the evidence moved the prevailing view from his steady state theory to the big bang theory. Once I began to come across ideas of multiverses, as a science-fiction enthusiast I was delighted, but as a science writer I often felt that there was way too much speculation and not enough testable science. Thanks to the many cosmologists, astrophysicists and science writers who have informed and frustrated me in the discovery (you will find details of some in the Further Reading section). Thanks also to the team at Icon Books, including Connor Stait and Steve Burdett for their ever-helpful input.

CONTENTS

ENTER THE MULTIVERSE 1

Multiverses turn up so often these days in films that it would be easy to think that the concept behind them was either a perfectly normal part of accepted reality, or a pure fantasy on the same level as superheroes[*] or magic powers. Marvel's 'cinematic universe' has been riddled with multiverses for years, while other movies such as the 1998 *Sliding Doors* take a more subtle approach of using a 'what if?' decision point to examine how each of two hypothetical parallel worlds would develop.

This is a small-scale version of the sub-genre of science fiction known as alternate history, where writers speculate on how the present would be if, for example, the Reformation never happened, or Germany had won

---

[*] Superhero stories are often presented as being science fiction – and this is occasionally feasible with a character like Batman or Iron Man, where there are no superpowers involved. But as soon as powers crop up there tends to be little more than a veneer of pseudoscientific explanation, pasted over what is simply magic.

the Second World War. Although in such fiction there is only the 'counter-factual' version of reality and any comparison with our world is left to the reader, there are still effectively two instances of the universe in play.

What is a multiverse?

Linguistically, the concept of a multiverse is more than a little baffling, as the universe is supposed to be the word used to describe all that there ever was or will be – the whole of existence. The word 'multiverse' seems to have been coined by the American philosopher William James who, in an 1895 contribution to the *International Journal of Ethics,* wrote, 'Visible nature is all plasticity and indifference, a multiverse, as one might call it, not a universe.' As is, perhaps, not atypical with philosophy, this is one of those sentences where it's possible to understand what all the words mean without having a clue what the writer intended. He certainly had nothing in mind resembling our modern concept.

James may have been suggesting that the universe did not have a single guiding force and meaning but rather was a chaotic environment, something reinforced by the physicist Oliver Lodge who in 1904 was reported as saying, 'The only possible alternative [to a universe with a purpose] was to regard the universe as a result of random chance and capricious disorder, not a cosmos or universe at all, but rather a "multiverse".' This philosophical nicety seems to rather overload the term 'universe' with more meaning than most endow it.

It is frequently said that as early as 1848 Edgar Allen Poe wrote about there being a number of universes – a multiverse by any other name. I suspect most who suggest this have not bothered to work their way through Poe's extremely long and turgid essay based on a lecture, *Eureka: A Prose Poem*. This is probably better known for presenting the correct solution to Olbers' paradox, which we explore further in the next chapter.

What Poe actually says is this: 'Have we, or have we not, an analogical right to the inference that this perceptible Universe – that this cluster of clusters – is but one of *a series* of clusters of clusters, the rest of which are invisible through distance …' He is making the distinction between what we now call the visible universe and the rest of it, containing plenty more galaxies (his clusters) – but he was not suggesting there was more than one universe.

In the modern physical sense, the word 'multiverse' seems to have entered the lexicon in a story by the English science-fiction and fantasy writer Michael Moorcock in 1963. His novella *The Blood Red Game*, published in the SF magazine *Science Fiction Adventures*, introduced the kind of multiverse familiar from comics and their movie spinoffs. (The comics had featured the concept earlier, but it was Moorcock who named it.)

As American physicist Brian Greene has pointed out, our definition of what a universe and a multiverse is can be decidedly vague. He recommends adopting 'the approach famously adopted by [US] Justice Potter Stewart to define pornography. While the U.S. Supreme Court struggled to delineate a standard, Stewart declared "I know it when I see it."'

Extra dimensions

For centuries, the size of the universe and whether or not it was finite fascinated scientists and philosophers. And that's part of the story. But however big it may be, once mathematicians and scientists began to think outside of the apparent natural limits of three spatial dimensions, it seemed possible that the universe we experience might not be all there is to reality.

An early addition to those familiar three dimensions was when time began to be considered as a fourth, if rather different, dimension. In his 1895 novel *The Time Machine*, H. G. Wells has his time traveller explain that: 'Clearly any real body must have extension in four directions: it must have length, breadth, thickness, and – duration.' This concept would become truly scientific with the development of Einstein's theories of relativity, which pull time and space into a single integrated body of spacetime.

At the same time, mathematicians were escaping from the familiar geometry that had reigned supreme since the work of Ancient Greek mathematician Euclid (and is still the most widely taught). Basic geometry is limited to two-dimensional, flat planes. Clearly this isn't enough to deal with the three-dimensional world around us. As a trivial example, we are taught that the angles of a triangle add up to 180 degrees. But you can happily draw a triangle featuring three right angles (270 degrees) on a sphere like the Earth.*

---

* The easiest way to envisage a three-right-angle triangle is to take two lines at right angles to each other down from the North or South pole to the equator, which they will meet at right angles.

That is only the start. We might experience three spatial dimensions, but mathematics is not limited to describing the physical world that we inhabit. In a mathematical universe there could be as many dimensions as were needed for the most extravagant flights of fancy. We might not be able to envisage, say, the true appearance of the four-dimensional hypercube known as a tesseract – but maths can easily describe it.*

Although the mathematical universe only partially intersects with reality, it can be an effective starting point for what-if speculation. That term 'what-if?' is arguably the basis of all storytelling, reaching its heights in science fiction and fantasy. As human beings we are not limited to what we have directly experienced – we can speculate and fantasise about possibilities that might never be susceptible to scientific probing.

Fiction writers picked up on the potential for adding dimensions early on. Arguably, though, the first text to play with dimensions involved removing a dimension rather than adding new ones. The 1884 book *Flatland* by the entertainingly named Edwin Abbott Abbott described experiences of creatures living in a

---

* The simplest three-dimensional projection of a tesseract can be envisaged as a pair of cubes, one within the other, with lines linking equivalent vertices, though all of the three-dimensional 'sides' formed this way are actually identical cubes. American science-fiction writer Robert Heinlein wrote a very impressive 1941 short story, *And He Built a Crooked House*, in which an architect creates a house in the form of an unfolded tesseract which, as a result of an earthquake, collapses into its true four-dimensional form.

two-dimensional universe and how they would experience three-dimensional objects passing through their plane. This gives us insights into how we might experience higher dimensions – but the far more common approach in fiction would be to extend the dimensional structure available to the writer. In a fantastical sense, this had been done for centuries, with magic employed to connect our world to a parallel location – be it a fairy realm or an astral plane.

Such imaginings were often described as featuring 'another dimension' or a 'parallel dimension' in a hand-waving sense, without really explaining what this meant. In reality 'another dimension' probably refers to the potential existence of another *three* spatial dimensions that don't intersect with our own at all, while a parallel dimension had the feel of three new dimensions parallel to our own but slightly shifted in yet another orthogonal direction. To make stories interesting, there usually had to be some kind of portal to bridge between the multidimensional spaces, though if considering science rather than magic as the mechanism, it was hard to see how such a connection could ever be made.

A good example of such an extra-dimensional approach is in Algernon Blackwood's 1917 story 'A Victim of Higher Space'. Likely inspired by Abbott, Blackwood imagined that physical objects we experience do have more than three spatial dimensions. So, for example, a box in our world would actually be (at least) four dimensional, we just can't experience the fourth. Except (by magical means, though wrapped up in 'psychic' waffle) one individual starts to experience this and can even move in the fourth dimension, which he can handle, but is terrified of slipping into even more dimensions and never returning.

One of the first stories to attempt a more scientific approach made use of the fourth physical dimension, time, rather than space as the dividing feature between two sets of universes. Murray Leinster's 1934 story 'Sideways in Time' effectively imagined a second time dimension running at right angles to the time we experience – so identical physical dimensions could be experienced with no intersection in time with our own world. More commonly, though, the new 'dimensions' would be in the same time stream but spatially separated, even if time was able to flow at different rates in the locations. This happens in everything from C. S. Lewis' *Narnia* to the 'demon dimension' of *Buffy the Vampire Slayer*.

Most of these examples, including the comic-book concept of 'parallel dimensions', have been based on pure fantasy with little more than magic to enable the transition between the two, but one of the most common 'cheat conventions' of science fiction also makes use of what amounts to extra dimensions. To make it possible to explore space in a timescale accessible to humans, writers often make use of some kind of faster-than-light drive. A common mechanism for this to work is to invoke a 'hyperspace' which is (somehow) dimensionally separate from everyday space. Movement in hyperspace works on a different scale to normal space, enabling rapid travel between two points that are far apart in the real universe. Not surprisingly, there is no known physical equivalent in reality.

Philosophical multiverses

But leaving aside speculation about invisible parallel dimensions, just having a single universe would eventually give astronomers and cosmologists a headache. This is because of something known as fine tuning: there are many constants of nature that would only have to be slightly different for life never to be possible. This seems so unlikely to happen by chance that many cosmologists, striving to avoid the need to invoke a creator, believe that there are a vast range of different universes, forming a multiverse with many variants of the universal contents, making ours just one of many that happens to be viable for life.

Then there is the argument from similars. Some astronomers (who perhaps should know better) have pointed out that a number of astronomical features we have historically thought unique – the Earth, the solar system, the galaxy – have turned out to be just one of many. So, they argue, this should apply to the universe too. This is, unfortunately, magical thinking, which is why I used the expression 'argument from similars', the pseudoscientific justification used for homeopathy. The idea in that unscientific discipline is that a treatment will be effective in homeopathic form if it has a similar effect to the disease or ailment it is treating.[*]

---

[*] The idea that homeopathy deploys poisons with similar effects to diseases and ailments would be disconcerting were it not for the way that the active substance is diluted so much that it is likely there is not a single molecule of it left in the final remedy.

Using such an argument to justify multiple universes seems an odd one to deploy with any scientific rigour: just because some things in science have turned out to be more numerous than was first thought does not mean all will. After all, early atomic theory assumed each type of substance had a different atom – so, for example, cheese atoms were different from bread atoms. In reality, our picture of fundamental particles has not multiplied as we discovered more but has greatly reduced – a frequently occurring situation that provides entirely the opposite reasoning.

Another challenge that cosmologists face is the mystery of inflation, the sudden vast expansion that the early universe may have gone through. Some theories predict this has to happen all the time, producing a whole range of unconnected universes. And then there are the vast numbers of parallel universes suggested by the 'many worlds' interpretation of quantum theory. To make it even more fun, some suggest that the cosmologists expecting a multiverse have got their understanding of probability all wrong.

This is all entertaining speculation, but while most agree there can be no communication between universes, some scientists have devised ways this might happen – or even expect universes to collide in extra dimensions, creating big bangs like the one we consider to be the start-point of our own universe. Welcome to one of the weirdest and most exciting aspects of modern physics and cosmology, attempting to explore the mysteries of what we can know of the outer limits.

Before we start to dig into the potential reality of the different forms of multiverse, let's take a brief visit to one way that, even in conventional space and time, the picture of a simple universe with three spatial dimensions at right angles to each other ceases to make sense. This is when we encounter the space-mangling form of a black hole.

Into the black hole

'Black hole' is a term that now appears regularly in common usage. However, the physical phenomenon described as a black hole is anything but something we come across in everyday life. In reality, because of the immense distances of interstellar space, humanity may never directly be able to experience a black hole, as happens all too often in SF movies such as *Interstellar* (and probably this is just as well). The concept of a black hole emerges from two aspects of physics – Einstein's general theory of relativity and the life cycle of stars.

To get his masterpiece to work, Einstein had to move away from the rigid three-dimensional geometry that dates in part back to Ancient Greece. He realised that massive bodies distort the space and time around them. For example, the Earth is moving in a straight line through space. But the space around the Sun is warped by the star's powerful gravitational pull sufficiently to turn that straight line into a curve, locking the Earth into an orbit. The planet's path through space doesn't twist – it is space itself that is distorted to produce the near-circular path.

When Einstein first published his theory, although he was able to specify the equations that described the actions of gravity, they proved too complex to solve. But another German physicist, Karl Schwarzschild, in hospital after being injured in the First World War, simplified the equations by imagining a very simple body that was entirely spherical, uniform and did not rotate. He discovered that if such a body were dense enough, its gravitational pull would distort spacetime sufficiently nearby it that light (or anything else moving) would have its path bent back into the body and could never escape – it was a theoretical description of a black hole. (The term itself was only dreamed up in the 1960s.*)

At the time Schwarzschild wrote his paper, in 1915, no one thought of this as anything other than a model which made it possible to explore one small aspect of Einstein's equations. But as more became known about the mechanisms driving stars it was realised that at the end of their lives, some very specific types of star would suffer a catastrophic collapse. As they ran out of fuel, resulting in no nuclear reactions to push against gravity, the most massive stars would collapse – and with sufficient mass involved, no force of nature could prevent that collapse from occurring. The star should

---

* The term 'black hole' is often attributed to American physicist John Wheeler, who certainly popularised it in the 1960s. The first recorded use of the term, though, was at a January 1964 American Association for the Advancement of Science meeting. Unfortunately, it was not recorded who used the term, but the article mentioning it in *Science News-Letter* was written by Ann Ewing.

effectively disappear to a dimensionless point known as a singularity.

Such a black hole would not appear as a point when seen from outside – from Scharzschild's solution it was clear that, for any particular mass, there would be a sphere known as an event horizon from inside which nothing, not even light, could escape. But the horizon is just that – not a membrane or barrier, but simply a distance from the centre of the sphere. An object passing through it would not be aware of doing so. Instead, it would head towards oblivion in the centre of the black hole, where the gravitational force should become infinite.

An important word there is 'should' – in practice, when anything physical becomes infinite (with the possible exception of the universe) the theory supporting it breaks down. We know that there are many bodies out there in the depths of space that we call black holes, and which have every appearance of being just that. But exactly what happens at the heart of the black hole is unknown – it seems likely that something would prevent singularities forming, but we don't know for certain.

What we do know, though, is that space and time distort more and more as something heads towards the centre of such a body. As a result, the familiar dimensions of time and space become so distorted that, in effect, a body falling to the centre of a black hole is heading for a point in time, not a point in space. The part of the universe inside the black hole is so different from the rest of existence that it could be argued it exists in a different universe in its own right – as we will see later, some have even speculated that our own universe is contained in the event horizon of a black hole.

But we are heading too quickly down one of the many conceptual rabbit holes that litter our view of reality. Let's take a step back and look at what we mean by the universe in the first place, and how ideas about it have evolved over time.

UNIVERSAL SPECULATION 2

Our modern vision of the whole of existence operates on several levels. With the exception of a few space flights, human life has been limited to the surface of planet Earth. This planet forms part of our solar system, based on our mid-sized star, the Sun. Our star is just one of over 100 billion in our reasonably chunky galaxy, the Milky Way. Although there are various nominal structures of which the galaxy forms a part, it is itself a small component in the observable universe, which is estimated to contain hundreds of billions of galaxies (we'll come back to the concept of 'observable' in a moment). And if there is such a thing as a multiverse it's possible that our universe is just one of many, but let's start with the assumption that the word means what it was historically understood to describe – all of space and time.

What is our universe?

The word 'universe' itself is something of an oddity. It means in the original Latin form *universum* 'one turn', though from the start it was taken to refer to everything that exists. It came to English via the French word *univers* in the late-sixteenth century. Some think that the 'turn' part implies that all the separate parts of existence are 'turned into' one thing by this word, though it seems a weak way to derive the meaning. Whatever, we are stuck with it. But the key part is that opening 'uni' – there should only be one universe in the way that the term was originally envisaged. If there is indeed a multiverse, then that strictly should be the universe, where what we can experience is just a small subset of the whole.

How small a subset our direct experience applies to becomes clear when we look at the limits of human exploration. At the time of writing, the human beings who are furthest from the surface of the Earth are the temporary inhabitants of the International Space Station. Although this is often described as being in space, which makes it sound extremely far away, the space station orbits at a ridiculously short distance from the rest of us – a little over 400 kilometres (250 miles) up. That is, for example, closer than Paris is to London or about the same distance as Washington DC to Pittsburgh.

The furthest that humans have ever travelled out into space is to the Moon. This means covering a more respectable distance of 384,000 kilometres (239,000 miles).

But that's around 160 million times closer than the nearest star other than the Sun, a red dwarf called Proxima Centauri, itself more than 10 billion times closer than the furthest distance potentially visible in the universe. The word 'universe' simply doesn't give a clear picture of just how enormous our environment is.

Even if there genuinely is only one universe, there is likely to be plenty of it that we can't ever know about – and this is where the word 'observable' comes in. Our ability to explore the universe beyond the near vicinity of Earth makes use of light, or strictly electromagnetic radiation in all its forms, including radio, microwaves, infrared, ultraviolet, X-rays and gamma rays[*] as well as the visible. Such radiation takes time to reach us, travelling at just under 300 million metres (a little over 186,000 miles) per second. This means that the more distance there is between Earth and an object, the further back in time we see it. The time it takes the light to reach us delays what we see, providing a window into the past.

So, for example, the nearest big neighbour to the Milky Way is the Andromeda galaxy, visible to the naked eye on a clear, dark night. This is around 2.5 million light years distant, so is seen as it was 2.5 million years ago. The equivalence between the numbers is because the measure of a light year is conveniently the distance that light covers in a year when travelling through the vacuum of space. In principle (though highly unlikely in practice),

---

[*] We can also make use of particles other than photons, such as neutrinos, to make some astronomical observations, but electromagnetic methods dominate.

the Andromeda galaxy could have disappeared entirely, yet we wouldn't notice it for those 2.5 million years.

The consensus is that the universe we can observe is around 13.8 billion years old. This would seem to imply that we shouldn't be able to see further than 13.8 billion light years away in any direction, as the light would not have been able to reach us if travelling from a greater distance. In practice, though, we have to modify this measure because during its lifetime the universe has expanded greatly. This means that the most distant objects that are potentially possible to observe are now around 46 billion light years away.

What there is beyond that horizon is unknown. While there are plenty of theories, there is no means to determine which is correct. The universe could be infinite. It could be finite but unbounded (more on what this means soon) or it could be finite and bounded – though minimum estimates suggest it is likely to be at least a hundred times bigger than the observable universe. Each of these concepts has its own appeal.

Big bangs and infinite universes

There is one paradox worth exploring here that emerges from the idea that the universe is 13.8 billion years old – how we come to the idea that the big bang, the beginning of the universe as we know it, happened around 13.8 billion years ago. This is deduced by looking at the rate at which different objects in the universe are moving away from each other, and how this has changed over time. The universe is not just expanding, but that expansion is accelerating.

If we imagine running the clock backwards, so everything was closer and closer to each other, there would come a point where everything in the observable universe was co-located, effectively in a single point. This means, incidentally, that we can pick any point in observable space and say 'that's where the big bang happened' because at the time all those points were the same place. You are currently located where the big bang happened.

Now we get to the mind-boggling bit. This would still be possible if the universe were infinite. When we describe the universe as expanding it is space itself that gets bigger – it is not like an object that is expanding in space. So, when we imagine running the clock backwards, it's not that we should think of the universe getting smaller in space, but rather that the distance between objects gets smaller and smaller. This is perfectly possible in a universe that was always infinite (and still is). Everything gets closer to its neighbours. The density of any particular volume of space increases to the state where, at the beginning of time, it reaches infinity and our theories fall apart – but the universe remains infinite.

If this still seems impossible, think of the infinite set of whole numbers along an infinitely long number line. Let's imagine there's a metre between each number. Now chop out all the even numbers and the space either side of them. We are left with the infinite set of odd numbers. There are still an infinite set of numbers along an infinitely long line – even though the 'distance' between each number has halved. If we continue this process indefinitely, any particular number is pretty much on top of any other … but the number line is still infinite in length. One of the problems we have with thinking about infinity is that our normal approach to arithmetic falls apart, making it difficult to visualise the result of making such changes.

Early infinite universes

The distorted idea of medieval science that many hold now suggests that the only viewpoint before modern science was introduced was of a flat Earth floating in a tiny universe no bigger than the solar system – but this couldn't be further from the truth. Firstly, while it's true that a flat Earth picture may well have been held by the uneducated (a viewpoint that still exists among a few today), our home planet has been known to be roughly spherical since Ancient Greek times.

Admittedly, the Earth looks fairly flat if you ignore hills and valleys, but its shape was determined by what we'd now regard as scientific reasoning, as it was reasonably easy to do experiments to demonstrate its curvature. However, much of the wisdom of the Classical era was based more on philosophical argument than experimental observation – more like the work of a court of law than science. The problem with this was that when, for example, arguing that the universe was infinite (containing far more than that 'just the solar system' picture) the result can feel more like sophistry than science.

A typical argument from Ancient Greek infinity supporters was that the definition of the universe was everything that existed. If the universe were finite, then it would have boundaries. And having boundaries means there is a dividing line between something and everything else. Unfortunately, if there was an 'everything else' then there was an inherent contradiction in the definition of universe. This meant that the universe had to be unbounded. As we have seen, it is in fact possible to be finite and not have boundaries in the conventional sense – but this is a discussion about terminology, not what is and is not in reality.

Before science began to be studied in the modern sense there were many attempts to deal with the scale of the universe. A useful way of getting a feel for the approaches taken was in the work of the remarkable Roger Bacon. A Franciscan friar living in the thirteenth century, Bacon wanted to write what amounted to an encyclopaedia, collecting together the knowledge of the day. This would be based not on pure philosophy (or theology), but on observation and mathematics.

Due to various restrictions that were placed on his religious order, Bacon needed permission from the Pope to write his book, and he set out to write an outline proposal. In doing so, he got carried away – his initial attempt turned into a 500,000-word volume, while two more drafts of a proposal resulted in a pair of extra volumes giving an extraordinary view of what was considered knowledge about the world in the 1260s.

That first volume, known as the *Opus Majus*, considered both the shape of the universe and its size. He considered egg-shaped, lentil-shaped, cheese-shaped* and angular forms, but dismissed these because when they rotated, they would leave behind a vacuum, which he followed Aristotle in believing could not possibly occur in nature.† This pointed him towards a spherical universe. (Thankfully, Bacon didn't think of the implications of moving the universe sideways.)

Bacon would also not accept the possibility of there being multiple universes, because again this would not result in an overall shape that was symmetrical under rotation, as he wasn't thinking of putting these universes in extra dimensions. Note, incidentally, that there was no concern about having Biblical support for his views, as would be deployed in countering the idea that the Sun doesn't move through the sky – these were purely philosophical observations based on what was a largely accepted view about the nature of a vacuum.

As for the size of Bacon's universe, he decided it wasn't infinite because of a (flawed) geometric argument. He imagined two lines heading from the centre of the universe outwards for ever. These would be the same length. He then imagined a third line starting partway up one of the original lines, parallel to the second original line.

---

* Bacon had in mind wheels of cheese, rather than wedges.

† The traditional rendering of Aristotle's view was 'nature abhors a vacuum'.

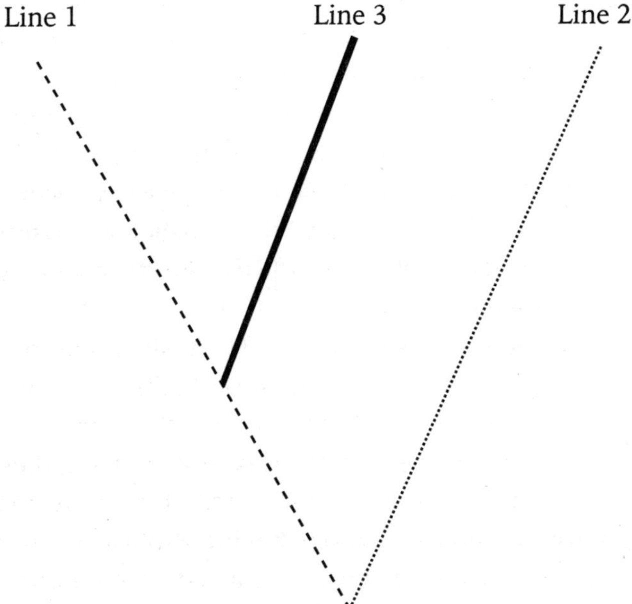

Bacon's argument against the infinite universe

This new line (darker in the figure above), he argued, would be the same length as the original lines, as all are infinite in length. This would make the new line the same length as part of the first original line and all of it – which he believed to be an impossibility. Sadly, as others would discover, it's not so easy to make pronouncements about infinity, or to base arguments on it.

What is infinity, anyway?

The reason Bacon got this wrong requires us to dip a little into the nature of infinity, a concept that is important when dealing with the universe. The Ancient Greeks had a concept of infinity that they referred to as apeiron, a word suggestive of chaos. Aristotle had helpfully pointed out what he regarded was the key aspect of infinity – that it was a potential entity, not one that was directly accessible.

He illustrated what it meant to be potential by using the example of the Olympic Games. Did the Olympic Games exist? Absolutely. But unless you are at a certain location at a certain time, he wasn't able to show you the Olympic Games. They were potential, rather than present and tangible. This 'potential infinity' proved very useful mathematically – it is essentially what's meant by infinity when it's used in calculus, for example. It is what is represented by the lemniscate symbol ∞. But it wasn't any use when thinking about a truly infinite universe.

It wasn't until Galileo took a diversion into infinity in his masterpiece *Two New Sciences*,* over 350 years after Bacon wrote his books, that more detail was cast on the mathematical behaviour of infinity. What Bacon hadn't grasped was that you can add or take away any finite number from infinity and the result is still infinite. Galileo played around with this in a number of ways, showing that infinity isn't a number in the normal sense. He made Bacon's argument useless because it is entirely possible to chop the finite end off an infinite line and still have an infinite line.

---

* Despite the hype, *Two New Sciences* made a far bigger contribution to science than Galileo's infamous book comparing the Copernican, sun-centred universe with Aristotle's Earth-centred system.

A later attempt to use logic to disprove the existence of an infinite universe proved harder to counter – and, as we have seen, strangely its flaw would be pointed out not by a mathematician or scientist, but by the writer Edgar Allen Poe. What has become known as Olbers' paradox was named after eighteenth-century German medical doctor and amateur astronomer Heinrich Olbers, who made it widely known. It points out that if the universe is infinite and the stars are distributed randomly, then looking in any direction should result in looking directly at a star. The entire night sky should consist of an infinite set of points of light. There should be no darkness between the stars.

Remarkably, despite Poe coming up with the correct counterargument, the paradox was still being incorrectly answered in the 1980s. In 1987 a study showed that 70 per cent of textbooks (not even popular science titles) gave the wrong explanation, with the lack of stars being blamed on interstellar dust or red shift. (This shift reflects the fact that the light from stars moving away from the observer will have a longer wavelength and lower energy. If this were to happen sufficiently, the visible light from the stars would no longer be in the range we can see ... though, of course, light that had too short a wavelength and too high an energy to be seen would move into the visible range.)

Poe's explanation in *Eureka* points out that because light travels at a finite speed, if the universe were extremely large, most of the light wouldn't have got here yet. He didn't think that the universe was infinite, but his argument was essentially correct. If the universe were infinite and uniformly filled with stars, but had not existed for infinite time, then the vast majority of the light heading in our direction would not yet have made it to us.

Finite but unbounded

The reality is that we can't know how big the universe is, and though we aren't aware of anything physical that is infinite, arguably the universe is one thing that could be … because what would stop it being so? There certainly is an attraction to living in an infinite universe, because it deals with the problem that went back to the Ancient Greeks. If it's finite, what happens at the edge? If you were in a spaceship and reached the edge, what would happen if you continued in a straight line?

One of the possibilities at the heart of the general theory of relativity is that space (and time) can be curved. In principle a finite universe could be curved in such a way that it does not have boundaries – the case mentioned earlier of it being finite but unbounded.

This may sound impossible, but we are very familiar with the concept of something being finite but unbounded when considering a two-dimensional surface existing in three-dimensional space. Think, for instance, of the surface of the Earth. It is definitely finite: travel around its equator and you will come back to your starting point in around 40,000 kilometres (25,000 miles). You can reach any point from any other after travelling a finite distance that is no more than half this distance. But the Earth's surface has no boundary, no ending. You can travel around it for ever without coming to the edge.

Similarly, a three-dimensional space could in principle be warped through another dimension so that any attempt to leave it results in re-entering from the other side. Getting to this viewpoint for a three-dimensional space is more of a stretch of the imagination – but it is mathematically possible (and bear in mind that the general theory of relativity makes it clear that space can be curved as much as we like). You could envisage what might happen if you travel outwards from Earth for far enough: you would eventually come back into the universe from the 'other side', never having left it – even though the higher-dimensional manipulation required to make this happen is beyond direct imagining.

This all presumes a single universe. One relatively simple form of multiverse can again be imagined by starting in two dimensions. Think of a tray filled with flat soap bubbles, leaving no gaps. In a finite tray, any bubble could expand if other bubbles contracted. Alternatively, in an infinite tray, all the bubbles could expand without any contraction. If we extend each bubble to three dimensions the same could apply to a multiverse.

We will come back to a multiverse based on bubble-like structures when we've established a potential scientific basis for it. But first we need to consider an even simpler way of bringing a multiverse into being. By going virtual.

VIRTUALLY UNIVERSAL 3

Before getting into the crunchier aspects of cosmology and quantum physics that give us concepts such as eternal inflation, bubble universes and the many-worlds interpretation, it makes sense to take a look at what is arguably the most scientifically justifiable theory, even though this is one that is generally rejected by most scientists. It all starts with stories.

Unreal universes

As I point out in my book *Brainjacking*,* one of the key factors distinguishing the human species from other animals is that we are storytellers. Story is central to our being. We use it to inform, influence and manipulate others all

---

* *Brainjacking: The Science of Influence and Manipulation* (Icon Books, 2024)

the time. Some of those stories are factual. That is (hopefully) the case with the stories in this book, though the stories that many of the featured scientists dream up are not always based on observation of anything we can detect in the real world. But, from the earliest days, many of our stories have been fictional.

A fiction writer has the ability to generate what could be considered an alternate world – or even universe. We even apply the term freely to films or TV shows. So, we might see a reference to the 'Marvel Cinematic Universe' for a movie that fits into Marvel's big picture linking different characters, or an entry into 'the Buffyverse' for a series that picks up on the particular rules of the universe that are special to the *Buffy the Vampire Slayer* TV series. Although this has always been the case with, for example, book series that have continuing characters, it has a much greater impact with science fiction or fantasy, where it's not just the individuals involved who are special to the series, but physical concepts might be set differently to those of the real universe. The *Star Trek* universe, for example, gives us faster-than-light travel, time travel and alien worlds with suspiciously human-like inhabitants, while fantasy usually brings magic, breaking physical laws with abandon, into the world.

It could be argued that these alternates provide us with a virtual form of multiverse – that we accept the existence of other universes, at least while we suspend disbelief and enjoy the film, TV show or book. But the approach is taken to the next level in video games, especially when the level of immersion in the gameplay is increased by deploying virtual reality. As yet it's not possible, but if that immersion could be taken to the point of being indistinguishable from the real world, we take one step closer to a multiverse made real through the virtual environment.

This possibility comes across clearly at the end of the 1999 movie *The Matrix*. A fair amount of the film's storyline depicts the central character, Neo, beginning to increasingly bend and then break the laws of nature in the virtual world. This climaxes in the triumphant ending of the first film in the series, where he can dodge bullets and fly. To all intents and purposes, for Neo the virtual world of the Matrix has become an alternate universe with new physical laws.

So far, we have been thinking about two distinct levels of universe. The physical universe we occupy, and the imaginary universe of the film, book or immersive virtual reality. But some scientists and philosophers believe we need to take a step further and consider the possibility that the existence we experience is itself virtual – and if that is the case, then we can sensibly consider the existence of such virtual universes as forming a true multiverse.

Living in a simulation

It's a big leap, but just imagine that there exists a civilisation with incredibly high technology capabilities, able to construct information technology as far in advance of ours as twenty-first century tech is beyond cave paintings. It's possible that such a civilisation could produce virtual worlds in their equivalent of computers. Worlds so sophisticated that virtual humans running in that software believed that they were real, inhabiting a real universe. If this were the case – and we'll go on to look at why some do believe this – then at the very least there would be a two-universe multiverse. There's the virtual universe we inhabit, and the physical universe occupied by that incredibly high-tech civilisation. But, of course, there's no reason why there would need to be just one virtual universe. There could be millions of them.

Some even use a probabilistic method to suggest we almost certainly do live in a virtual universe. The argument goes that if we assume there may only be one physical universe suitable for life, but there are millions of virtual universes, then the chances are that we live in a virtual universe. From the viewpoint of safely using probability, this is a dubious argument (the problems are similar to those of the fine-tuning argument we will go into in more depth in chapter five, which is taken more seriously, so I won't go into detail here). Yet if the technology were practical to be able to construct a virtual universe, then it certainly would at least be a possibility that we were part of one.

At first glance, though, our universe seems far too complex to be virtual. It's not just a matter of the billions of people on our planet and their intricate daily interactions. As we've already seen, the universe is massive in scale. The Milky Way alone contains at least 100 billion stars – and there are estimated to be billions or even trillions of galaxies in the visible universe. To simulate all of this would surely require computing power beyond even the most advanced of civilisations.

It has also been pointed out that if there were a fully detailed simulation of the universe, it would also include the ability for entities within the simulation to produce a simulated universe. And the people (or equivalent) in those simulations would also be able to produce simulated sub-universes. And so on. It's a picture that's reminiscent of the English mathematician Augustus De Morgan's humorous verse, 'Big fleas have little fleas upon their backs to bite 'em, and little fleas have lesser fleas, and so ad infinitum.' Except, this level of complexity wouldn't really be necessary to make virtual universes possible.

Simulating consciousness

At its simplest, a universe simulator could have just one fully rounded persona in it – you. Simulating even one person's conscious mind is currently well beyond our capabilities. Despite the impressive abilities of generative AI such as ChatGPT, the software has no understanding or reasoning ability. It is merely pattern-matching prompts to data that is based on content already produced by human minds. The reason that an AI can make such silly mistakes is that lack of reasoning.

As an example of how easily it can go wrong, I recently put a query into the search engine Google, asking a question to help research something related to the inflation concept that we will cover in chapter seven. Specifically, I asked, 'Why was the BICEP2 telescope built at the South Pole?' It's quite difficult to uncover the reasoning behind undertaking the considerable effort required to construct a complex structure at one of the most inaccessible places on Earth. At the time of writing, Google has recently started to lead its response to a search with an AI-generated summary. In this case it told me:

> The BICEP2 telescope was located at the South Pole to observe the cosmic microwave background in a clear, dark, and cold environment. This helped the telescope detect primordial gravitational waves, which could provide evidence of a theory of inflation.

It is certainly a good guess that you might want to locate a telescope at the South Pole because it's dark there. There is no sunlight for around six months of the year, and of course there are no streetlights or any of the other light sources of modern life that so plague astronomers. The problem is, though, that the (now replaced) BICEP2 was a radio telescope working in the microwave region – and a lack of background signals in this part of the electromagnetic spectrum is not described as being dark.

But the real disaster in the confident result produced by the AI comes in that final sentence. The location did not help the telescope to detect primordial gravitational waves. Admittedly, the team behind it did announce that they had done so in 2014 ... but then they had to withdraw the claim within weeks when their discovery turned out to be the result of polarisation of the microwaves when they came into contact with dust. That final sentence is pure fiction.

Despite the over-excited claims of some AI enthusiasts, we are not yet close to being able to produce artificial general intelligence, capable of the same kind of reasoning as humans. Even if and when this becomes possible, this doesn't necessarily mean that the software would experience consciousness as we do. But it isn't inconceivable that our imagined advanced civilisation could do this.

Limiting simulation

With the virtual you in place as the only conscious entity, only the parts of the universe you have direct exposure to need any depth of simulation. Your close human contacts will need relatively complex simulations, but don't need to have any 'life' outside of their interactions with you. As for anything outside of the lights you can see in the sky, the chances are that everything you know about the rest of the universe is the result of indirect observation. All that vast quantity of supposed existence doesn't need to be simulated at all.

The only argument that can make this picture of a virtual universe seem rather silly is that what you have heard about the universe (if only in reading this book) is unnecessarily complicated. If you were the only conscious part of the simulation, why bother with all those details? It would seem enough to have a single city, or at best a planet with a few decorative extras. But that assumes we can attribute to the creators of the virtual universe the kind of reasoning we might apply. They might simply be having fun, creating as complex a universe as they are able to, in part, perhaps, to give you something to wonder about.

Like much of the speculative science you will meet in this book, it is arguable that the concept of a virtual universe is impossible to prove or disprove – it is likely to be at best 'ascientific', meaning that it isn't accessible to science. Admittedly there has been plenty of science fiction suggesting ways that we might detect limitations in the simulated universe which indicate its nature. This is rather like the denouement of the 1998 film *The Truman Show*, where the central character Truman Burbank discovers he is living in a (much simpler) non-virtual simulated world when his boat hits the edge of the 'sea' he is sailing across. But, as yet, we have been unable to detect anything that would indicate such a flaw in reality. (Or, at least, that's what the simulation programmers want you to think.)

For the rest of the book, we will stick to the assumption that if there is a multiverse, it has physical reality. Such a viewpoint starts with the possibility of the existence of more than the familiar four dimensions: three of space and one of time. We need to get dimensionally complex.

DIMENSIONAL COMPLEXITY

4

As we have seen, early fictional nods to the multiverse concept often referenced 'another dimension'. In the physical world we have traditionally thought of space as consisting of three spatial directions at right angles to each other. While purely mathematical dimensions certainly can be considered to add in extra imaginary orthogonal directions, such dimensions can equally be thought of as a set of numbers that can go from having no interaction at all, to having complex rules as to how they interact.

Setting things straight

We start with sets. Although it was a relatively late innovation in our mathematical thinking, the set is a fundamental structure of mathematics, and set theory underlies much of the basics of arithmetic. A set is a group of things.* You could have, for instance, the set of people called Brian, or the set of things that are orange. In mathematics, sets can be used to split up the numbers – for example, the set of the odd numbers, or the set of the prime numbers.

One aspect of set theory that has escaped into common usage is the term 'subset' – this describes a set that is also part of another set, illustrating that sets can be nested inside each other. Take, for instance, the integers (whole numbers). The set of the odd integers is a subset of the entire set of the integers.

In familiar space and time, we can define the location of any object by a set of four values – the object's position along each of three spatial dimensions at right angles to each other, plus its location on the time dimension – in effect the point in time when we are observing the object. There need not be any relation between the elements of a set (as its components are called) other than what each describes. So, for example, we can add a fifth element to the set describing that object's location that is its colour, a sixth that is its name, and so on.

Some of the elements of the set may be totally distinct – but others have the potential to interact. And it isn't always the most obvious connection that produces a relationship. For example, the three spatial dimensions, despite all being of the same kind, would normally be

---

* A set is a group in the normal usage of that word. Confusingly, mathematicians use 'group' to mean a particular type of set where there is a specifically defined relationship between every part of the set and some of the other elements.

considered independent. But if the object is moving, the values of the object's location in those dimensions will change with time. Similarly, the colour might change with time – for example, if the object fades in sunlight. And the name might change depending on the language spoken in the physical location that the spatial dimensions indicate.

Time and relative dimensions in space

Although, as we have seen, Wells explicitly referred to time as a fourth dimension – and a different position in time can be linked to a different position in space – time was historically thought of as something physically separate from space. A different kind of measure. However, soon after Einstein came up with his special theory of relativity, the German mathematician Hermann Minkowski (who had taught maths to Einstein) realised that the theory implied that space and time were so fundamentally linked that they could not sensibly be considered as two separate entities. Instead, Minkowski proposed a combined dimensional set of spacetime.

Einstein himself was initially highly sceptical of this step, in part because of the way that the concept of extra dimensions had been popularised at the time. His concern arose from the development of the spiritualist movement. All religions that promise a life after death other than reincarnation tend to have some kind of 'other universe' concept for what happens to those who have died. But spiritualism, at its peak of popularity in Einstein's lifetime, treated the afterlife differently.

This movement represented a merger of religious and scientific thinking. It was based on assuming that life after death was not a supernatural concept, as is the case with heaven or the equivalent posited by some religions, but rather involved a soul moving into a different 'plane' – effectively a parallel set of physical dimensions where they were no longer present in our universe but in an 'elsewhere', from which it was possible to communicate through spirit mediums.

Just as some modern thinkers pick up on a vague understanding of scientific concepts such as 'quantum' and 'vibrations' and apply them without any scientific evidence or logic to their supposed healing applications, many in the early part of the twentieth century, aware to some degree of the new scientific and mathematical thinking of the day, considered extra dimensions to be a useful prop for their ideas. This is why when Einstein arrived in New York in 1930 and was asked by a reporter about the fourth dimension, he apparently first replied 'ask a spiritualist'.

But Einstein was persuaded of the value of spacetime because of the impact that his special theory has on the passage of time in moving bodies. When, say, a spaceship is moving relative to the Earth, then the passage of time on that spaceship as seen from Earth slows down. (This is true of every moving object, but it only becomes clearly noticeable with something moving as fast as a spaceship can.) Take this to the extreme in the so-called 'twin paradox' and you can imagine a space traveller returning to Earth aged forty, arriving just in time for her identical twin sister's seventieth birthday.

This linkage between space and time, which has been proven time and again by experiment, becomes even clearer with Einstein's second venture into relativity, his general theory, which shows how massive bodies warp spacetime, producing the effect that we describe as gravity. Not only does gravity slow the passage of time, the mechanism for an object starting to fall will only work if space and time are merged.

An image often used to help understand how general relativity results in gravitational effects is to imagine the straight-line path of an object moving through space as a line marked on a flat rubber sheet. We now put a bowling ball on the sheet near that path. This warps the rubber, causing the straight line to bend around the ball. It is this distortion of space by a massive object like a star or planet that cause satellites to orbit. However, this analogy is useless to understand why a stationary object released near a massive object will begin to fall.

It might seem that the object would naturally move towards the planet because it would roll down the rubber sheet. But what would make it roll? We are imagining it rolling as a result of gravity – and we can't use gravity to explain gravity. In reality, what happens is that the object, though stationary when measured on spatial dimensions, is moving through the time dimension. And the distortion of spacetime caused by the planet's mass causes the change in time to be warped into a change in space. There is a direct crossover between the space and time dimensions in as simple an action as an apple falling.

Extra dimensions

Once we have the picture of space and time as our familiar set of four dimensions, we can picture what may be possible if we add in further dimensions. One clear option, as we saw in the first chapter, is to have either a single extra time dimension – so objects can exist in the same space but without impinging on our reality's time dimension – or to have three additional spatial dimensions, which would mean an object based on these new dimensions would never interact with anything in our three spatial dimensions.

Such a parallel universe would never be detectable from our own. But adding in a single extra universe is just a starting point – bear in mind that dimensions are mathematically just a set that can have as many elements as we wish. Why stop after adding just one or three extra dimensions? There could be 20, 2,000 … even an infinite set of dimensions that we would never experience.

This isn't entirely satisfactory as explanatory science in the sense that without any interaction at all there is no difference to anything we can experience between a simple, single universe and this collection of parallel (strictly, orthogonal) dimensions. We wouldn't see anything different. In fiction, though, the other dimensions are usually accessed via some sort of 'gateway' or 'portal'. In the vast majority of cases, the mechanism involved to make this happen is effectively magic. There is no scientific basis for it. However, in principle we could imagine some kind of dimensional warp that would make this possible.

Imagine a (theoretical) object that is truly one dimensional – a straight line in space that has no thickness at all. We can rotate that object through 90 degrees, and it will have moved from one dimension (say horizontal) to another (let's call it vertical). Of course, any physical object we can directly experience has some breadth and height, not just length, so it can't be said to only exist in one dimension. Even so, this gives a picture of the potential for warping an object into a new set of dimensions.

It might seem that the obvious parallel here is the way that the general theory of relativity tells us that massive objects warp spacetime. After all, this involves a distortion in reality from one dimension to another. But the problem here is that the massive object causing the warp already exists in the four dimensions of our universe. To warp some extra dimensions into our own would require a massive object that could somehow straddle the two universes. Later on, we will discover the theoretical ekpyrotic multiverse (see chapter nine), which envisages a mechanism for interaction between dimensionally separated universes, but without that approach it is hard to see how this can be possible in reality.

This doesn't mean that extra dimensions have not been postulated for our universe – in one sense, this is an approach that has proved very useful since the concept of imaginary numbers was introduced all the way back in the sixteenth century. We often credit small children with dreaming up special imaginary friends – this is an imaginary friend of mathematicians that has proved an extremely valuable aid to physicists and engineers.

Imaginary dimensions

Getting imaginary all begins with the rather more familiar idea of negative numbers. Unlike the positive numbers, we don't directly experience negatives – in fact, the positive numbers are sometimes referred to as the natural numbers for this reason. I can easily show you 3 oranges,[*] but I can't show you −3 oranges. A negative value is in effect a mathematical metaphor for an action taken to make a positive number smaller. If we add our −3 oranges to 5 'real' oranges we end up with 2 oranges because we have taken 3 away.

In the basic manipulations of arithmetic, the negative number is reasonably well behaved. But it throws up a problem if you attempt to take its square root. The square root of a number is the value which, when multiplied by itself, results in the number required. So, for example, the square root of 4 is 2, because $2 \times 2 = 4$. What, then, is the square root of −4? It's clearly not 2. But equally it's not −2, because when we multiply −2 by itself, the result is also equal to 4. Any positive number has two square roots – in this case both 2 and −2 are square roots of 4.

Of course, this restriction is only a problem in the real world – just as mathematicians can make use of as many dimensions as they like, they can dream up numbers with whatever properties they like, as long as they behave consistently in response to the rules of the mathematical system of which they form a part. The concept of the

---

[*] Conventionally we refer to numbers smaller than either 10 or 12 as words, but these are particularly mathematical oranges, which need a numerical representation.

square root of a negative number was first mentioned by the Milanese mathematician Girolamo Cardano, who in 1545 wrote a book on solutions to algebraic equations.

In his book, *Artis Magnae Sive de Regulis Algebraicis Liber Unus* (book one of the great art of algebraic rules), Cardano briefly discusses the solution to the equation $x^2 + 1 = 0$. If we take 1 from either side of the equals sign, this is the same as $x^2 = -1$, so x is the square root of -1. Cardano didn't do anything with this other than commenting that the solution was 'as subtle as it is useless'. To stress the dubious nature of such values, French philosopher René Descartes dismissed them as being 'imaginary numbers'. This resulted in the square root of -1 being represented by the symbol i – but such numbers would only be a philosophical entertainment until they were used to open up an extra dimension.

A common mathematical concept is the number line. It's usually thought of as a horizontal line with zero in the middle, stretching off to infinity in both negative and positive directions. Arithmetical operations all involve movement up and down the number line. Until the nineteenth century, it had always been conceived as a single entity. But then the German mathematician Carl Gauss realised that it could be considered a dimension – and the nature of imaginary numbers meant that they could be used as values on a second number line, representing a second dimension. This could usefully be regarded as lying at right angles to the familiar number line.

Now, instead of a number line, there was a two-dimensional number plane. Any location on the original number line could be considered 1 multiplied by any positive or negative value. Similarly, any location on the imaginary number line was represented by i multiplied by a positive or negative value. And just as the familiar Cartesian coordinates* represented a point on a plane using x and y values, so any position on the number plane could be represented by the combination of real numbers and imaginary numbers, shown as, for example, $3.5 + 2.7i$, known as a compound number.

It might feel initially as if Gauss was just reinventing Descartes' wheel – but imaginary numbers have a relationship with real numbers that fits perfectly with the geometry of flat surfaces, while x and y coordinates are a pair of unconnected numbers. Because of the way that imaginary numbers multiplied together produce real numbers, complex numbers could be used with all the tools of algebra that had been established over the centuries. Complex numbers proved to be ideal vehicles for working on waves, for example, which can be considered two-dimensional in nature and are used widely in everything from basic engineering to quantum mechanics.

Quaternions and more

Once dimensions had infiltrated arithmetic and geometry it was almost inevitable that attempts would be made to incorporate more than two of them. In the 1840s, Irish mathematician William Hamilton extended complex

---

* Cartesian coordinates are named after their creator René Descartes, though using two numbers to identify a location on a map was, for example, mentioned by Roger Bacon.

numbers to form quaternions, picking up on an earlier term for a collection of four poems. Quaternions in mathematics have the form $a + bi + cj + dk$ – where i, j and k are three different orthogonal sets of imaginary numbers, taking the whole viewpoint to four dimensions. This was necessary to cope with changes over time in three spatial dimensions.

Quaternions were nowhere near as easy to handle as complex numbers. Apart from anything they are described as noncommutative: for two quaternions $q1$ and $q2$, $q1 \times q2$ is not the same as $q2 \times q1$. But they can be manipulated using variants of the algebraic methods applied to basic compound numbers. Hamilton was convinced that quaternions would transform applied mathematics and spent the rest of his career working on them. In practice, their use proved relatively limited, but they continue to be employed for some types of movement calculations in three-dimensional space, notably in high-end computer graphics.

Hamilton had hoped that the four-dimensional powers of quaternions would be particularly useful in dealing with interaction between vectors. These mathematical extensions of a single number typically combine two values – both the size and the direction of a physical quantity. For example, where speed is a single value (a 'scalar'), just the measure of the rate at which something moves, velocity is a vector because it combines speed with the direction of movement. Hamilton correctly realised that the four values of the quaternion would enable him to work on the interaction between vectors that were changing – but the approach was clumsy.

It was his contemporary, Scottish physicist James Clerk Maxwell, who would be among the first to build on

Hamilton's imaginative idea and develop vector calculus, which does for vectors what basic differentiation and integration does for ordinary numbers. Maxwell came up with the terms 'convergence', 'gradient' and 'curl' to represent different forms of quaternion operation which would be carried through into vector calculus, though 'convergence' was replaced by its opposite 'divergence', and the longer names shortened, ending up with the operations div, grad and curl, still widely used in physics.

Extra dimensions, then, had helped power up applied mathematics – but these were still operating in a mathematical world, removed from the reality of the universe (or multiverse). Despite Einstein's resistance to extra dimensions, from 1919 onwards he would experience suggestions that extra *real-world* dimensions existed on top of the familiar four, notably in the work of the German mathematical physicist Theodor Kaluza. In a letter to Einstein, Kaluza suggested adding a fifth dimension to spacetime. This dimension would not be directly detectable, but was, Kaluza believed, a mechanism that would enable electromagnetism to be brought into the same theoretical sphere as gravity by incorporating electromagnetism into Einstein's field equations of general relativity.

Invisible dimensions

It's worth taking a moment to observe the elegance of Einstein's equations as they are usually represented:

$$G_{\mu\nu} + \Lambda g_{\mu\nu} = (8\pi G/c^4)\, T_{\mu\nu}$$

What's shown on the page looks deceptively simple – but getting to this in 1915 had required Einstein to stretch his mathematical capabilities to the limit (and occasionally beyond). It was arguable that achieving the general theory of relativity was what was on Einstein's mind in 1943 when nine-year-old Barbara wrote to him complaining about her difficulties with maths. Einstein replied: 'Don't worry about experiencing difficulties with maths, I can assure you that my own problems are even more serious!'

The reason that the apparently straightforward formula above is anything but simple (and also why what appears to be a single equation is referred to in the plural) is because each of the elements with the subscript $\mu\nu$ is in reality a complex mathematical construct called a tensor. These are like vectors on steroids – each tensor can represent a multidimensional set of equations in its own right. In the case of the general theory, the tensors collapse ten complex differential equations* into a single symbol.

Einstein's tensors are four dimensional to deal with the workings of gravitation on spacetime. But Kaluza added a fifth dimension to the tensors to allow them to

---

* A differential equation shows the relationship between a function and its derivatives. A function is a mathematical structure that converts one set of numbers into another set of numbers – so a simple function could be 'multiply every number by 2' – but it could also be an extremely complex set of operations. Derivatives measure the way that one variable changes as another does – so, for example, how speed changes with time. Derivates are found using the mathematical method of differentiation.

cope with electromagnetism as well. Not only did this give the equations enough capacity to deal with the requirements of electromagnetism, Kaluza found that they enabled Einstein's formulation to produce Maxwell's equations, the definitive description of how electromagnetism acts. It felt like a triumph.

It's worth briefly mentioning why this was considered so important – and why many scientists have spent their careers trying to unify forces of nature. We generally consider there to be four fundamental forces. The ones we experience directly are gravity and electromagnetism, which is not just an explanation of electricity and magnetism, but also why, for instance, you can sit on a chair rather than your atoms passing straight through it. The other two are the strong nuclear force, which hold particles together in the atomic nucleus, and the weak nuclear force, which is responsible for some changes to atomic nuclei.

While it is entirely possible that these forces could have been totally unrelated, it has seemed likely to physicists for over a hundred years that the forces were related and could be described by a combined theory. Quantum physics and particle theory have brought together three since Kaluza's day – but there is a fundamental incompatibility between gravity and the other forces. It does not appear to have the 'granular' quantum nature of the other three. So had Kaluza been able to combine gravity and electromagnetism with a single set of equations it would have been remarkable.

Note, before we head back into Kaluza's fifth dimension, that just because mathematicians and physicists use more than three or four dimensions does not necessarily mean that they think of these dimensions as being a part of the framework of reality, in the same way that we typically do for the three spatial dimensions and time. It is often convenient to use multiple dimensions mathematically as a representation of a range of different varying values that are not directly equivalent. One such dimension can, indeed, represent a potential location in space or time. But just like the extra elements in a set mentioned earlier, it could equally refer to temperature, or magnetic field strength, or any measure that can have a range of possible values, like how sleepy you are feeling.

Kaluza was envisaging something very different, though, not unlike an extra spatial dimension. He dealt with the observation that nothing suggested the existence of such a dimension – one had never been detected – by requiring the fifth to be quite different, making it unmeasurable. This was referred to as the 'cylinder condition' because Kaluza thought of the dimension having a kind of rotating existence outside of the familiar fixed framework.

Although Kaluza's extra dimension had achieved the remarkable feat of combining gravitation and electromagnetism into the same set of equations, it proved a fragile concept that did not reflect what is observed in reality. If a theory is to adequately describe the universe (or multiverse), or even part of that whole, it must be capable of surviving changes of approaches to viewpoint and measurement, whether it be rotating the system or moving to different coordinates.

Coordinate systems provide a shorthand way to pinpoint a location on one of more dimensions. On the number line, the coordinate system is the collection of numbers that line represents. We can refer to coordinates in two dimensions by a whole range of ways. The familiar Cartesian coordinates, giving measures on the x and y axes, for example. Or, as we have seen, compound numbers can be used, combining a real and an imaginary number. Equally, the distance from a particular point in the two-dimensional space combined with an angle of rotation from a starting direction can identify a location. As we increase the number of dimensions we can always find a range of ways of specifying a position – and to work physically, a model needs to cope with changing the coordinate system, something that Kaluza's five-dimensional model was incapable of delivering.

For that matter, as Einstein would later point out, there was an arbitrariness to the nature of the five-dimensional space. Why was one dimension spinning about, meaning it had to be treated so differently to the rest? A second theoretician, the Swedish physicist Oskar Klein, would produce a similar five-dimensional model, but one where the fifth dimension was somehow rolled up to be incredibly small, an alternative way of making it undetectable. Their combined work, known as Kaluza–Klein theory, would form the basis of string theory when the extra forces of nature were added in, requiring a whole collection of curled-up dimensions.

As we will discover in chapter nine, the development of string theory opened up new possibilities for variants of the multiverse, but would continue to be plagued with problems and arbitrariness. Before we get to that, though, we need to explore one of the main arguments used to support the need for a multiverse to exist in the first place. Our universe seems incredibly well adjusted for life. It appears as if it is fine-tuned for us to be able to exist – something that might delight theologians, but horrify many scientists who weren't prepared to consider the existence of a creator.

FINE TUNING CONUNDRUM 5

By now, it is becoming clear that multiverses might be fun to speculate about, but they have lacked a scientific justification. Any thoughts about their existence were based on abstract theory, rather than observation or experiment. However, cosmologists in the second half of the twentieth century began to favour the possibility of the existence of a multiverse not so much because there was any evidence that they exist, but rather because they didn't like the implication of a multiverse *not* existing. At the heart of this concern was the concept of fine-tuning.

A Goldilocks universe

When we look at the details of physical reality, a surprising number of physical constants* have to be pretty much

---

* Physical constants are measures of various aspects of nature that are usually considered unchanging, such as the charge on an electron, the speed of light and so on.

what they are for life to exist at all. If, for example, the strengths of the electromagnetic or gravitational forces were significantly different we would not expect solar systems to form and hence life as we know it could not exist. Similarly, the strength of the strong nuclear force, which is central to the formation of atoms and hence matter, balances on a knife edge. It has to be in a very narrow 'Goldilocks' region that is not too strong and not too weak.

As England's Astronomer Royal Martin Rees has pointed out in his book *Just Six Numbers*, if we look at a measure called the efficiency release of energy when hydrogen fuses in stars to make helium, which gives us a good measure of the strength of the strong nuclear force, it can be given a value of 0.7 per cent. If that value were instead 0.6 per cent or less, the only atom that could form would be hydrogen. If it were 0.8 per cent or more, hydrogen would be practically non-existent, meaning no stars, no water and no organic molecules to be the building blocks of life. For that matter, if the 'cosmological constant', a value that describes the strength of dark energy, the force behind the expansion of the universe, were almost imperceptibly larger, stars would not have a chance to form.

To give another example, if neutrons, the particles with no electrical charge that form part of almost all atomic nuclei, were less than 1 per cent lighter, atoms couldn't form.* This is because the other particle type of

---

* To be precise, rather than 'if neutrons were 1 per cent lighter atoms couldn't form', it is strictly if the quarks, the particles that make the neutrons up, were 1 per cent lighter.

the nucleus, the positively charged protons, would decay into a neutron and a positron (a positively charged equivalent of an electron). Without protons, there would be no atoms. There are at least a couple of dozen such 'finely tuned' values that seem suspiciously well fitted to the existence of stars, planets and life. It's almost as if things were designed this way, a phrase that is itself designed to seriously worry most scientists.

Going anthropic

The pragmatic approach to our fine-tuned reality is what is sometimes called the weak anthropic principle. This is true, but facile. It says that if the constants of nature were inimical to the existence of life, we wouldn't be here. The term was first introduced in 1973 by Australian astrophysicist Brandon Carter. There's a useful philosophical view in cosmology known as the Copernican principle, stating that there's nothing special about how things look from Earth. Carter suggested that this needed a slight modification, because effectively our existence put some limitations on nature. (As we have found out more about exoplanets, the Copernican principle has suffered another slight modification, as it has become apparent that Earth-like planets appear to be fairly unusual.)

In a sense, the weak anthropic principle certainly holds true: as we are here, then the constants of nature must be those that support us being here. Full stop. But this feels unsatisfactory from a statistical viewpoint. The problem is that it is apparently highly unlikely that so many such values would all occur randomly. Surely, it is argued, there must be some reason for it. It has been estimated that the chances of the universe being fine-tuned enough for life is around 1 in 10^{136} (that's 1 followed by 136 zeroes). To put it another way, the likelihood of things being as they are in our universe is not dissimilar to the chances of winning a national-scale lottery around twenty times in a row.

Those who believe in a creator of the universe have no problem with this, but it has resulted in some interesting reflection from atheist scientists (almost certainly in the majority), such as the late Stephen Hawking. In his widely purchased, if less widely read, book *A Brief History of Time*, Hawking wrote:

> The remarkable fact is that the values of [fundamental constants] seem to have been very finely adjusted to make possible the development of life. For example, if the electric charge of the electron had been only slightly different, stars either would have been unable to burn hydrogen and helium or else they would not have exploded ... One can take this either as evidence of a divine purpose in Creation and the choice of the laws of science or as support of the strong anthropic principle.

This strong anthropic principle is a beefed-up equivalent of the weak version that explains our extraordinary luck in being able to exist by positing the existence of a multiverse. It requires either a whole set of coexisting universes, or a single universe that is somehow partitioned into different non-interacting regions (which is, to all intents and purposes, still a multiverse). The idea is that each of these universes or regions can have different fundamental constants. As a result, it is suggested, it becomes clear how the universe can have constants so fine-tuned for life: because the vast majority of universes don't. If a tiny percentage of them have the right values, then we can invoke the weak principle and say that it is only in these universe(s) where life will inevitably exist.

Hawking goes on to cast doubts on the strong anthropic principle, though clearly cannot support the alternative, and presumably was hoping for a third option. Another great English physicist, Fred Hoyle, had a similar problem as an outspoken atheist. Hoyle was one of my scientific heroes growing up. Knowing his views on religion, I was surprised to read him quoted as saying:

> A common sense interpretation of the facts suggests that a superintendent has monkeyed with the physics, as well as chemistry and biology, and that there are no blind forces worth speaking of in nature. I do not believe that any physicist who examined the evidence could fail to draw the inference that the laws of nuclear physics have been deliberately designed with regard to the consequences they produce inside stars.

Dig a little deeper and we discover that Hoyle never said this – the paragraph is a merger of two different quotes from him. The first half was from a 1980s essay,*

---

* *The Universe: Past and Present Reflections*

while the second was taken from the transcript of a talk he gave in Cambridge in 1959.* To be clear, Hoyle was not supporting any established religion, because he rightly believed that there was an incompatibility between the rigidity of some aspects of traditional religious belief and the inherent flexibility of good science. He commented, 'There is a clear reason for this. All formal religions were devised at earlier times when man's understanding of the physical world was far less developed than our present understanding. It is natural therefore that modern science should find itself at odds with these earlier attempts at an expression of the religious impulse of man.'

Hoyle bridged this distinction between believing in a formal religion and his conviction that there should be some sense of design more explicitly than Hawking, saying:

It may surprise you when I say that I have yet to meet a person who was not imbued by a religious sense. The great difference between us lies in our varying attitude to formal religion. Religion in a non-formal sense I take to mean that a man will look up at the stars at night with a sense of awe, that he will feel that the majestic play of the universe has some deep-laid purpose, and that his own small role in the play must make sense, if only he has the wit to find it.

---

* Collected in *Religion and the Scientists*

Avoiding creation

Many modern scientists (unlike the vast majority of their predecessors) are atheists and so are resistant to the idea of some form of creator. Their viewpoint can be traced back in particular to the work of Pierre-Simon Laplace. In the early years of the nineteenth century, this French natural philosopher made clear his belief in a Newtonian clockwork universe, remarking:

> An intellect which at any given moment knew all of the forces that animate nature and the mutual positions of the beings that compose it, if this intellect were vast enough to submit the data to analysis, could condense into a single formula the movement of the greatest bodies of the universe and that of the lightest atom; for such an intellect nothing could be uncertain and the future just like the past would be present before its eyes.

Such a deterministic universe, where each event followed on predictably from the last in mechanical fashion, seemed to do away with the need for a deity. Laplace is apocryphally credited with responding to Napoleon when asked how God fitted into his worldview: 'I have no need for that hypothesis.'

As it happens, Laplace's determinism would be shattered by quantum theory, which would show that many events in the universe – for example, the decay of a radioactive particle – are impossible to predict from past events. There is true randomness involved. However, this was not sufficient reason to move away from a scientific viewpoint that made it necessary to exclude the divine as a driving force behind detected phenomena. After all, even if you believed in God, this was a supernatural belief, and science had become the study of nature.[*]

This argument would take longer to resolve in the field of biology than astrophysics. The heavens, after all, seemed to be definitively like clockwork, providing a very obvious celestial timepiece, but living organisms appeared to require creation (hence the still commonly used term for living organisms, creatures). By far the best-known image that would be used to hold off the biologically minded atheists of Victorian times was Paley's watch.

This was an argument given in a book by the Victorian clergyman William Paley called *Natural Theology*. Paley noted:

In crossing a heath, suppose I pitched my foot against a stone, and were asked how the stone came to be there; I might possibly answer, that, for anything I knew to the contrary, it had lain there forever: nor would it perhaps be very easy to show the absurdity of this answer. But suppose I had found a watch upon the ground, and it should be inquired how the watch happened to be in that place; I should hardly think of the answer I had before given, that for anything I knew, the watch might have always been there. Yet why should not this answer serve for the watch, as well as for the stone? Why is it

---

[*] I say 'science had become the study of nature', as originally 'science' (*scientia* in Latin) was a more general concept similar to knowledge, which meant that theology was regarded as a science.

not as admissible in the second case as in the first? For this reason, and for no other, viz. that, when we come to inspect the watch, we perceive (what we could not discover in the stone) that its several parts are framed and put together for a purpose ... if the several parts had been differently shaped from what they are, of a different size from what they are, or placed after any other manner, or in any other order, than that in which they are placed, either no motion at all would have been carried on in the machine, or none which would have answered the use, that is now served by it.

Variants of this argument are still used by those who struggle to get their heads around the concept of evolution by natural selection. But for anyone with a grasp of science, Darwin's 'blind watchmaker', to use Richard Dawkins's evocative term, makes perfect sense and does away with the need for a purposeful creator to produce complex biological organisms.

With this background, it's not surprising that many cosmologists have stuck with the implication that fine-tuning must require the existence of a multiverse, because they are unwilling to explore the possibility of any form of purpose behind the existence and structure of the universe. The idea of the universe having a purpose is certainly ascientific – not susceptible to scientific argument – though the existence of a multiverse where we can't detect any other universes than our own suffers from the same problem. Interestingly, the best challenge to their logic comes from a profession that Stephen Hawking had previously derided. Philosophy.

Philosophical wisdom

In a 2010 book, *The Grand Design*, Hawking, with co-author Leonard Mlodinow, told us that there was no longer any need for philosophy because science was ready to answer all the big questions of life, the universe and everything. I thought the authors were wrong at the time, and a counter to the multiverse argument for fine-tuning presented by English philosopher Philip Goff underlines how valuable good philosophy (I am the first to admit that the field contains plenty of hot air – I'm not talking about *all* philosophy) can be.

As we have seen, the argument deriving a multiverse from fine-tuning goes something like this. Bearing in mind we are yet to discover any reason to explain why most physical constants have the particular values that they do, then if we think of every possible universe there could be, with every possible value of those various constants, it becomes vanishingly unlikely that the one and only universe would happen to have values that are capable of supporting not just matter, but life. Yet that's what we see when we take a look.

This fine-tuning appears to be incredibly unlikely. As we've seen, it's a bit like winning a lottery, but many times over and with only a single ticket ever being sold. However, people *do* win lotteries every week, because in reality there are very many tickets sold, making it much more likely that at least one of those tickets will be drawn. If we lived in a multiverse, where our universe was just one of a vast number of universes, each with different potential settings for the physical constants, then it's not particularly unlikely that at least one universe would have the right settings for life. We can then invoke the weak anthropic principle to say that the (small) set of such universes has to include our reality, otherwise we wouldn't be here to observe it.

This argument has always felt specious to me. To begin with, there's an assumption for which we have no evidence – that if a multiverse existed, then its different component universes would have different values of physical constants. But why assume that? And even if the multiverse explanation were valid (which, as we will see, it isn't), as Goff points out, this doesn't rule out alternatives. Some kind of purpose in the universe (divine or otherwise) is certainly also an explanation, and if we were to take the multiverse argument seriously, its proponents can't dismiss such alternatives unless they can provide an explanation as to why this particular untestable option is not to be preferred to their own.

More importantly, the probabilistic argument used seemed to be misusing mathematics, but I could not put my finger on exactly how it was doing so. All I could reason was that *any* specific universe was unlikely, with our particular version being no less likely than any other – so it simply happened to be the one that came up. Unlikely things happen. Goff, though, has a beautifully clear description of why the argument simply doesn't hold up in something known as the inverse gambler's fallacy. (This is not his invention, being already well-known in probability circles, but many cosmologists seem not to have discovered it.)

The inverse gambler's fallacy

The starting point is the gambler's fallacy. This may be familiar to you already if you have read anything about probability (see *Dice World* in Further Reading for more on probability). It is logically obvious that this fallacy is wrong, yet it remains painfully difficult to reject because it feels so natural. Imagine, for example, you repeatedly spin a fair coin. Each time for ten spins in a row, the coin comes up heads. We know that over time the coin is likely to come up with increasingly similar numbers of heads and tails. So, surely, after all those heads, it's more likely that the next throw will be a tail? Yet logic tells us that this can't be true. The coin has no memory. It doesn't know what happened before. This means that the next toss will still have a 50:50 chance of heads or tails coming up – and so it continues. The same independence of outcomes applies to any gambling process relying on random selection.

To see the inverse version of this fallacy, imagine you walk into a casino (it's no coincidence I am using gambling analogies here – gambling and probability are, of course, intimately linked). At your side is a friend who happens to be a cosmologist, and the first player you see as you enter the building is winning time after time. Your friend says, 'Look at that, the casino must be full!' This seems a bit of an odd assertion based solely on what you are seeing, so you ask her why she thinks this. She says, 'If the casino were empty, it would be extremely unlikely that this person would be able to keep winning like that. But if the place is packed, it becomes a lot less improbable that someone would be on a winning streak.'

Unfortunately, the cosmologist's argument, which is exactly the same as that used to suggest there should be a multiverse because of fine-tuning, is just as wrong as the traditional gambler's fallacy. As we have seen, the next outcome in a sequence of coin tosses has no link to the ones that came before.* Each time, you are observing a single throw, unconnected to all the others. Similarly, in the inverse fallacy, you have only observed one person's success in the casino. There is no connection between their outcome and that of other players. You can't deduce anything about the rest of the people in the casino from what is happening to that single individual. Similarly, we know that cosmologists have only observed a single universe that appears to be fine-tuned – this means that they can't use this observation to deduce anything about other universes. It is a clear mathematical fallacy.

Just in case the cosmologist resorts to the anthropic principle to back up her assertion, Goff has an answer to this too. The principle might incline them to say, 'Ah, but we are bound to see a universe that is fine-tuned, because that's necessary for life to exist.' It's true, of course – but it is a powerless argument. Again, we have to ask how it can tell us anything about the existence or otherwise of further unseen universes. As Goff points out, all you need do is add in a sniper to the casino scenario.

---

* There is one circumstance where the next coin toss *is* related to the outcome from the previous toss, which I demonstrate when giving talks on probability: if you have a double-headed coin.

This sniper is aiming at you, the observer, and is positioned so that he can see the outcome of the play by the lucky winner before you can. If that so-far lucky player fails to have a good outcome, the sniper will shoot you. You will never see the player make a bad play – the sniper makes sure of this, just as the anthropic principle prevents you from existing in a universe where … you can't exist. But that has no impact on your cosmologist friend's ability to deduce anything about the casino's fullness from observing this player.

In case this all feels a bit hand-waving, Goff brings in the concept of 'the requirement for total evidence', or 'the principle for total evidence', which is a basic requirement to make a safe deduction from probability data, first identified by German-American philosopher Rudolf Carnap in 1947. This is often used in a legal setting, where using partial evidence presents a real danger for a mistrial. So, for example, if someone is discovered carrying a knife and this is used as evidence to show that they could be responsible for a stabbing, it would be inappropriate not to include the evidence that the knife in question was a blunt bamboo one that was given with a takeaway, totally incapable of causing a stab wound.

Goff points out that what is being used in the casino case to infer that the casino is busy is the evidence that 'someone in the casino has had an extraordinary run of luck', where the complete evidence is actually that the *one individual studied* has had an extraordinary run of luck. The same applies to our universe, as the only universe studied in the multiverse argument.

I can't deny that there are plenty of philosophers who talk tosh. But, personally, I'm very happy to disagree with the late Stephen Hawking and Leonard Mlodinow here. The fact that the fine-tuning hypothesis is still held up by many as evidence for a multiverse demonstrates that we still need a certain kind of philosophy. One that is supported by an understanding of probability. There are philosophers and philosophers – not all have a grasp of mathematics, but some certainly do.

If we set aside the argument from fine-tuning, it may seem that it is not going to be possible to deduce the existence of a multiverse from any logical argument. Yet there is one other possibility that involves bringing in the big guns. In fact, the biggest possible: infinity itself.

INFINITE POSSIBILITIES 6

As we have already seen, we don't know if the universe is finite or infinite. But some theoreticians suggest that if the universe were infinite, it would effectively force it to be a multiverse in its own right. Unfortunately, infinity is a subject that can easily trip up the unwary, so we will need to take a dive into one of mathematics' most mind-boggling subjects to be sure of whether or not this makes sense.

Measuring the infinite

A starting point is to realise, like Galileo, that infinity is not a number in the normal sense. It describes something without limit. To get our heads around the concept of infinity and how it could have the ability to conjure up a multiverse, we first need to go into a bit more detail in an area we've already touched on: the mathematics of sets. Unlike a number, a set can be infinite in size, but there's a problem in envisaging this, as we usually think of the size of something being described using numbers. However, the appropriate tool to describe the size of a set is its cardinality.

The useful thing about cardinality is that it can be used to compare the size of sets even if we don't (or can't) know how big they are. Two sets have the same cardinality if we can go all the way through one set, pairing off the items in the set with items in the other set. In my book *A Brief History of Infinity* I demonstrate this by pairing off legs on my dog with horsemen of the apocalypse. As I can pair each leg with a unique member of the set of horsemen, I can say the sets have the same cardinality, even if I didn't know how many legs or horsemen there were. (I do.)

Of course, with a pair of infinite sets it's not possible to do this in as explicit a way. But as long as I can set up a pattern, enabling me to work through the full set, pairing off members of each set, then I can say the sets have the same cardinality. This approach, sometimes known as one-to-one correspondence, can even be used to show that a set is infinite. An infinite set is one in which the set has the same cardinality as subsets of that set. This ability is going to prove important when we think about an infinite multiverse.

We can show such pairing is possible by considering two sets: the integers – the whole counting numbers – and the even integers. The integers provide an infinite set. If it were finite, then there would be a biggest value – let's call it X. But if that were the case, there's a bigger number: X+1. So there can't be a biggest number. The even integers are clearly a subset of the integers. Every even integer is a member of the set of integers, but there are plenty of other integers (the odd ones) that aren't in the even set. Now, let's imagine pairing off each integer with the even number that is twice its size. We can pair 1 with 2, 2 with 4, 3 with 6, 4 with 8 … and so on. Each integer has a doubled value which will be an even integer. This means that the integers have the same cardinality as a subset, the even integers.

We are going to use cardinality to take infinity to the next level, but let's start by seeing how it's possible to argue that if the universe is infinite, it follows that we live in a kind of multiverse. Imagine we have such infinite space available to play with. As we have seen, our visible universe is around 92 billion light years across. We could go even bigger and choose an observable universe that's 200 billion light years across. (The number we select doesn't really matter.)

Because our imagined universe is infinite, we can incorporate as many sub-universes as we like into it – all the way up to an infinite set of them. You might feel uncomfortable with the idea that we could fit an infinite set of sub-universes into our overall universe, but we can make use of the cardinality of infinite sets to see that this is perfectly feasible. Imagine the space taken up by each universe is N cubic metres. Then we can pair off N with 1, 2N with 2, 3N with 3 and so on. N times infinity has the same cardinality as infinity. No problem.

A multiverse in a single universe

We can, then, accommodate an infinite set of such sub-universes into the overall infinite universe. And because each visible universe can't see or interact in any way beyond its light boundary, we can treat each visible universe as a true universe in its own right. It might be a matter of playing with terminology, but it would then seem reasonable to call the whole infinite universe a multiverse, composed of a potentially infinite set of smaller, independent universes.

Those who believe that an infinite universe implies a multiverse of this kind point out an interesting consequence. In this model, each sub-universe is finite. But if you have an infinite set of finite universes, some of them must be identical. This is because there are only a finite number of ways of arranging every possible occupant of each point in space (whether those occupants be atoms or empty space). The number of arrangements is vast ... but not infinite.

As American physicist Brian Greene, an enthusiast for the infinite multiverse points out, just 52 cards in a pack can be arranged 80,658,175,170,943,878,571,660,6 36,856,403,766,975,289,505,440,883,277,824,000,000,0 00,000 different ways. The contents of every point in the space of a universe in this picture would have imaginably more ways to be arranged. But that number of ways remains finite. And infinity is bigger still – so some universes would have to be identical. This would become true as soon as you had more universes than the number of possible arrangements inside a universe. In reality, an

infinite multiverse would be much bigger than this still –
any particular arrangement of contents could crop up an
infinite set of times.

Interestingly, Greene argues something that could
be assumed to be subtly different: 'the arrangement of
particles within patches *must* be duplicated an infinite
number of times' (my emphasis). Greene's patches are
my sub-universes. In principle, the difference between
'could be' and 'must be' need not be incompatible, as I
was referring to a particular arrangement of a universe,
where Greene wasn't. But in practice things get a little
more complicated.

Greene then goes on to point out that his definition
of the contents of a universe is entirely physical – the
assumption here is that if we can duplicate the nature and
location of every particle in the universe, we have an identical universe. There is nothing more than the physical. As
he points out, this is a matter of choice, but it is the choice
the majority of scientists would make. However, he then
goes on to make a more dubious leap. He goes on to say
that 'if the universe is infinite in extent, you are not alone
in whatever reaction you are now having to this view of
reality. There are many perfect copies of you out there in
the cosmos, feeling exactly the same way. And there's no
way to say which is *really* you.'

This is where argument-from-infinity meets its first
problem. There is an assumption here that every possible configuration of the universe will be duplicated many
times over. But that does not follow. The infinite set of
universes argument does not ensure that all possible con-
figurations will be repeated; there could be a finite subset

of possible universes – even just one – repeated an infinite set of times.

Greene also suggests that, given enough time, the universes in the multiverse would start to overlap. In my arbitrary setting of universes that are 200 billion light years across, eventually light would be able to travel further and cross the boundary between universes, though exactly how and when this would work would also depend on how the multiverse or individual universes were expanding.

Hilbert's multiverse

What Greene's 'quilted multiverse' (his term) picture seems to miss is that in an infinite multiverse, each component universe does not have to be finite: each could also be infinite. This is perfectly possible in an infinite multiverse, as illustrated nicely by an imaginary construct known as Hilbert's Hotel.

This imaginary resort is named after the German mathematician David Hilbert who, among other achievements, nearly beat Einstein to publication of his general theory of relativity by picking up on an early error in Einstein's workings. Hilbert's Hotel is a destination with an infinite set of rooms. We are asked to imagine that, despite this, when you arrive to check in you discover that there is not a single empty room. You are just about to head off and find a different place to stay when the mathematically savvy receptionist points out that all they need to do is to move the guest from room 1 into room 2, the guest from room 2 into room 3 and so on. Each current guest will still have somewhere to stay, but now room 1 is freed up for your use.

At this point, a remarkable coach turns up with an infinite set of extra guests to accommodate. Surely the receptionist will be defeated? But no, he puts each current guest into a room with twice the value of their room number. The room 1 guest goes into room 2, the room 2 guest into room 4 and so on. Now the whole infinite set of odd-numbered rooms is empty and all the new guests can be accommodated. And we can continue to do this for an infinite set of times. We can incorporate an infinite set of infinite coaches into Hilbert's Hotel just as we can easily fit our infinite set of infinite universes into an infinite multiverse.

If each universe in the multiverse is infinite, then we have no repetition requirement. Each can be totally different (or, say, identically empty). Note also there's a 'wheels within wheels' possibility here. Every argument that was applied to the multiverse with finite sub-universes could be applied to each infinite sub-universe … and so on. This is more than mind boggling. It demonstrates that arguments from infinity are inevitably flawed because there is always another possibility. You might say there is an infinite set of them. With no mechanism for making a choice between them.

A bigger infinity

There is one last potential consideration to be made if we do think that the universe (whether singular or multiversal) is infinite. Not all infinities are the same size. This is a particularly challenging assertion, but it was cleverly demonstrated by the German mathematician Georg Cantor,* who was also responsible for much of the development of set theory.

We have, to date, been using the infinity of the integers as our basis for considering infinity, how different infinite sets relate to each other, and what this tells us about an infinite multiverse. But Cantor pointed out that it is an assumption that all infinite sets have the same cardinality, and he was able to prove that they didn't. The specific set we can make use of to demonstrate this is the set of all the numbers between 0 and 1. We are not now dealing with integers but rather with decimals.

Cantor's proof is more rigorous than the description I'm about to give, but it demonstrates the thinking behind it. Imagine writing a list of every number between 0 and 1. If it has the same cardinality as the integers, then it is still the same kind of infinity. To check this, we need to pair off the values between 0 and 1 with the integers. I can't list the first few here in numerical order, as they will all start 0.0000 … heading off towards infinity with just a single digit difference beyond our reach. So instead, let's scramble up the numbers. Our list might now start something like this:

---

* Cantor is sometimes said to have been driven mad by his contemplation of infinity. He certainly ended his life in a mental institution. However, the picture is rather more complex, and it is arguable that a major contributory factor was the way that better-known but less imaginative mathematicians ruined his career by attacking his ideas.

0.**7**9023861 …
0.0**4**539358 …
0.26**7**56197 …
0.869**8**6262 …
0.0095**7**293 …
0.47356**5**98 …
0.943442**9**7 …
0.3122613**2** …
0.85084360 …
0.84411339 …

…

All I would need to be able to do is match each number in this list with an integer and I have shown that the list of numbers has the same cardinality as the infinite set of integers. But before attempting this, I'm now going to generate a number using the first digit after the decimal point of the first number, the second digit of the second number and so on (highlighted in the list above):

0.74787592 …

Finally, I'm going to add 1 to each digit after the decimal place (if the value is currently 9, I will make it 0):

0.85898603 …

It might not seem it, but this is a very interesting number. The first digit is different from the first digit in the initial entry in the list above. The second digit is different from the second digit of the second item in the list above ... and so on. This number, starting 0.85898603 ..., is not in the list at all. It is impossible to construct a complete list of values between 0 and 1 that will have the same cardinality as the integers. This infinity, known as the infinity of the continuum, is bigger than the infinity of the integers.

This is amazing in its own right, but it can also feed into our thinking about an infinite universe/ multiverse. At the moment we have two incompatible physics theories at the heart of understanding the universe. The theory that deals with gravity – the general theory of relativity – is a continuous theory. The space and time it deals with is not 'quantised' or broken into minimally small chunks. But most other physical theories are quantised. As we have seen, vast amounts of effort have been put into trying to unite these theories, perhaps by constructing a quantum theory of gravity.

When we look at an infinite multiverse using the infinity of integers, we are, in effect, describing quantised spacetime. And it is very probable that this is what we need to work with. However, if it turned out that the general theory of relativity was not wrong in dealing with a continuum of possibilities – if there was no way to unify these theories because they are simply unconnected things – then perhaps those who like to think about the implications of infinity for cosmology should be working instead with the infinity of the continuum – making the potential infinite components even more complex.

Fun though playing with infinity is, it does not enable us to produce more than an argument that a multiverse could exist without any evidence to back it up. But as we find out more about the origins of the universe, it seems there is one theoretical event in the early life of the universe that could imply the existence of a multiverse. We are talking about inflation.

INFLATING EVERYTHING 7

One way that multiverse theories interact with the more basic aspects of cosmology is through a concept known as eternal inflation. But before we can take that on, we need to understand what cosmologists mean by inflation in the first place. It's quite a story, involving faster-than-light expansion of the universe a tiny fraction of a second after it came into existence.

A uniform universe

It's not always noticed how good physicists have been at stealing terms from other scientific disciplines. In some cases, they have been so successful that the physics version is the one that comes to mind when the general public are shown the term. Think, for instance, of 'nuclear' … we instantly jump to a source of power (or destruction) based on the workings of the inner parts of the atom – yet 'nucleus' and its derivatives were lifted straight from biology and the nucleus of the cell. In the same vein, 'fission' – again likely to generate immediate physics associations – was copied from the biologists' concept of cells splitting.

However, if asked about 'inflation', the immediate association will tend to be economics and how far your money will go. It's on the news practically every day. A more physical possibility is what we do to a car tyre. But for cosmologists, the term has a very different meaning, devised to deal with a serious problem in an early version of the big bang model known as the horizon problem.* From all we can tell about it, our universe appears to be too uniform. This seems crazy when you look into the night sky – it's anything but uniform. But this refers to a specific type of radiation that fills the universe.

As we have seen, although we believe that the universe is about 13.8 billion years old, and that it became transparent to light about 300,000 years after the big bang, we can see a lot further than 13.8 billion light years – potentially up to around 46 billion light years, whichever way we look. This is possible because the universe has expanded. Light that set off towards us a little under 13.8 billion years ago will give us a view of objects that are now 46 billion light years away.

We can take an overview of the universe from an all-pervasive 'glow' known as the cosmic microwave

---

* Inflation was actually devised to deal with a more esoteric problem involving magnetic monopoles, and later also the so-called flatness of the potentially curved space that makes up the universe, but from our viewpoint, we can limit our look at it to its relationship to the so-called horizon problem.

background radiation. This is light that started flowing across the universe when it became transparent, and that has been on the move ever since. Initially, this electromagnetic radiation would have consisted of ultra-high energy gamma rays, but thanks to the universe's expansion, it is now in the microwave region.[*]

The background radiation was first discovered in the 1960s, initially thought to be some form of interference rather than a real effect. Famously (or infamously), its 1965 discoverers, Robert Wilson and Arno Penzias, working at the Bell Labs Holmdel facility in New Jersey, wondered if the background hiss they picked up wherever they pointed their antenna might have been caused by 'white dielectric material' in the horn-shaped antenna. This was pigeon droppings. The pigeons were removed to a distant location, but unfortunately returned and had to be disposed of more permanently. But the radiation continued to be detected.

It was quickly realised that this was, instead, the remnants of the predicted radiation from the earliest transparency, still crossing the universe. A sequence of three satellites has brought us more and more details of the cosmic microwave background in the often-displayed full sky images, which are usually represented in a shape not unlike a flattened egg.

---

* Microwaves are high energy radio waves with wavelengths of between around 1 millimetre and 1 metre. They were originally primarily used for radar, until a microwave engineer discovered they had melted a chocolate bar in his pocket and the microwave oven was conceived.

Cosmic microwave background from the Planck satellite

This radiation is extremely uniform in energy. It doesn't look to be the case in the picture above because the amount of variation in levels is artificially exaggerated. There is only about a 1 in 100,000 difference in energy between the strongest, brightest values and the weakest dark ones. And it is this kind of uniformity that gives rise to the so-called horizon problem.

Often the way this is described is to think of a photon of light coming from the limits of visibility towards us from each of two opposite directions. How, we are asked, could those photons know what each other's 'temperature' is (the photon's energy in this context*) to be in

---

* Depending on their temperature, objects give off different colours of light (think, for example, of heating up a piece of metal in a forge). Because of this, although the vacuum of space doesn't have a temperature in the normal sense, we can speak of electromagnetic radiation with a particular photon energy having a related temperature. The temperature of the cosmic microwave background radiation is around 2.73 kelvin – that's 2.73 degrees above absolute zero, or around −270 degrees Celsius (−454 degrees Fahrenheit, though scientists would never use that measure).

equilibrium, when they are so far away from each other? After all, while there is nothing that prevents space from expanding faster than the speed of light, information can't travel faster than light through the universe.

Unfortunately, this way of describing the horizon problem causes very reasonable confusion. Because the whole idea of the big bang theory of the origin of the universe is that if we imagine the lifetime of the universe being run backwards, the whole thing gets smaller and smaller until we eventually reach a point (in space and time) where everything is superimposed. In this starting position, there is no distance for the information to travel. There is no distance between these now-distant bits of the universe.

The argument to counter this goes that the universe began expanding from an infinitesimal time after its beginning, not leaving enough time for the different parts of the universe to get into near-thermal uniformity. Temperature is a measure of the energy of the particles in something, and, assuming they don't all start off the same, it takes time for them to interact with each other and even out to a uniform temperature. Note, though, that there is still an assumption being made here – that there wasn't always thermal uniformity, other than small quantum fluctuations that resulted in the cosmic microwave background's variation. We don't know how the universe came into being, so it is impossible to say definitively what its constitution was at the moment of the big bang.

Pump up the volume

However, in the early 1980s the apparent challenge of the horizon problem looked in danger of putting the by then dominant big bang theory into question. A solution was proposed by American physicist Alan Guth, working at Stanford University. Guth suggested that the horizon problem could be alleviated if very soon after the big bang (we're talking around 10^{-32} seconds – 1/100,000,000,000, 000,000,000,000,000,000,000th of a second), the universe suddenly, and for no obvious reason, went through an extremely quick expansion … and then suddenly stopped expanding at this ridiculously quick rate and went back to its comparatively slow rate of growth.

In a similar fraction of a second to inflation's starting point after the big bang, the universe would have become 10^{26} or so times bigger than it had been (that's 1 followed by 26 zeroes), ending up with what would become the visible universe expanding to around three metres across, before inflation switched off again. It's an interesting idea, but one very much plucked out of the air, rather than being based on any evidence for this mechanism existing.

To be fair to Guth, he did devise inflation with a built-in reason behind its ceasing, down to the mysterious-sounding concept of a false vacuum. This is not some kind of fake Hoover, but a state of space that arises from a mechanism analogous to the concept of supercooling. Supercooling is the ability of something that should make a 'phase change' from liquid to solid, say, to be cooled beyond its freezing point while remaining liquid. This can happen, for example, if there are no impurities in the liquid for crystals to form. If a small object, acting as a seed for the transformation, is introduced it will suddenly and rapidly undergo a complete phase change.

When we think of a vacuum, what comes to mind is a volume of space which doesn't contain anything. Unlike the atmosphere of the Earth, for example, the vacuum of space has no air in it – and a perfect vacuum in a pre-quantum view of physics would contain no particles whatsoever, though in practice there are plenty flying around in many areas of space.

However, in quantum physics, the idea of a vacuum becomes rather more complicated. Here it refers to the lowest possible energy level of a system, sometimes called the zero-point energy. It might seem that this should be the same as having no energy, but one of the implications of the uncertainty principle in quantum physics is that the smaller a period of time we look at, the more uncertain the energy level of a system is.

Given this picture, a false vacuum is a state of a volume of space that appears to be vacuum state and can be temporarily stable, but can then lose energy and drop down to the true vacuum, often as a result of the input of energy. A useful analogy is a ledge on the side of a mountain. If a mountainside is smooth, a rock rolled from the top will end up at the bottom, minimising its potential energy and producing its maximum possible kinetic energy. But if there is a ledge partway down, the rock can stop when it reaches that point. Locally, this is the place with the minimum potential energy, but it is greater than the true minimum below. Give the rock a shove – add a small amount of energy to the system – and it will continue down to reach the true minimum.

Guth proposed that inflation was the result of a local false vacuum condition. In effect, the extra energy of this state would drive the rapid expansion as the region collapsed to a true vacuum, rather as supercooled water will rapidly turn to ice if a small impurity is introduced. How the false vacuum came into place is rather more obscure, but it was blamed on a concept called spontaneous symmetry breaking, which is thought to have occurred in the early universe as different fundamental forces started to break away from each other.

In Guth's model, the equivalent of the shove given to the rock was a quantum process known as tunnelling. This is the result of another strange aspect of quantum particles – their location is not absolute, but has a range of possible values until they interact with their environment, when they take on a specific value. This means that a quantum particle can appear on the other side of a barrier, effectively without passing through that barrier. Tunnelling is a widely observed phenomenon, and one that is used in electronic devices – for example, the flash memory in memory sticks (thumb drives) and solid-state disks rely on the ability of quantum particles to tunnel through an insulating layer and interact with the data inside.

To make inflation work, Guth imagined the energy in the form of particles held in the false vacuum tunnelling through the local barrier that keeps it in place, taking the energy down towards that of the true vacuum and releasing it to drive inflation. In the mountain ledge and rock analogy, tunnelling would involve digging a hole in the ledge so the rock can drop through. Like all analogies, though, we need to be aware that quantum tunnelling is not literal tunnelling. As we have seen, the particles don't pass through the barrier keeping them in place – they have a probability of being on the other side of the barrier without ever passing through it.

Unfortunately, quantum tunnelling proved an ineffective mechanism to be involved in inflation, because its probabilistic nature meant that it would result in a mix of false vacuum and the real thing, producing a much patchier result than that observed in the cosmic microwave background. An alternative approach got around this by using a different mechanism to release the energy – but this required a very specific fine-tuning of a special field* that had to be invented to make the release fit the timings which inflation required.

Looking for ripples

Although the idea of inflation continues to have strong support, it faces significant challenges which, if anything, have got worse rather than better over time. As a result, an increasing number of cosmologists and astrophysicists are looking for alternatives. One problem is that the detailed picture from the cosmic microwave background is not as supportive as a broad overview is. The background isn't as uniform as inflationary theory predicts it should be.

Specifically, this is about how measurements of the 'temperature' of different parts of the background

---

* This is a field in the physics sense – like the Higgs field that gives some of the particles their mass – not the more conventional meaning of an open space with grass.

radiation are distributed. The expectation from theory is that the variations should follow a normal distribution. This is a statistical pattern that is very common in nature – it is often depicted as a 'bell-shaped' curve – with a peak of occurrences around the middle, while gradually falling off into long tails on either side of the centre. The normal distribution describes the spread of everything from people's job satisfaction and test scores to sizes of pebbles on a beach. The cosmic background radiation in practice appears to have too many cold bits, distorting the distribution.

Another problem for inflation arises from our understanding of gravitational waves. These are ripples in spacetime itself, most commonly generated at measurable levels by the interaction of massive bodies – for example, when two black holes collide. Given that inflation's rapid start and finish would have provided a dramatic shakeup of the newly emerging universe, the expectation would be that it would have generated extremely powerful primordial gravitational waves. These would be too attenuated by the expansion of the universe that has happened since to be directly detectable (remember that after inflation the observable universe was only a few metres across), but should show up as a special kind of variation in the cosmic microwave background.

Because these ripples would still have been significant at the time the microwave background radiation was able to start crossing the universe, they should have an impact on the distribution of a measurable aspect of that radiation. And that's exactly what was announced to have been discovered in 2014, thanks to the experiment based at the

South Pole called BICEP2 (Background Imaging of Cosmic Extragalactic Polarisation) mentioned in chapter three.

This was a very specialist radio telescope that used 512 superconducting sensors* that measure the tiny amounts of energy gained from individual incoming photons making up the cosmic microwave background. The location at one of the most remote spots on Earth was chosen to minimise interference from manmade sources, inevitably a problem when working with radiation in the radio and microwave ranges.

As we have seen, the 'P' in the BICEP name refers to polarisation – a property of electromagnetic radiation such as light or microwaves that is based on the directions in which the electrical and magnetic variations that make up the radiation point, at right angles to its direction of travel. The expectation is that if those primordial gravitational waves from inflation existed, they would produce a particular curved pattern of polarisation in the cosmic microwave background described as a 'curlicue'. And this is the discovery that the experimentalists made public.

To be more precise, the process of announcement was a little more dubious than a simple announcement. The BICEP2 team had not announced the finding of polarisation attributed to primordial gravitational waves, when it first reached the scientific world in January 2014. It was

---

* Superconductors, such as the material used in the BICEP2 sensors, are materials in a special state that offers no electrical resistance, usually as a result of being cooled to very low temperatures. This can make it possible to produce sensors far more sensitive than normal microwave receivers.

leaked online (allegedly unintentionally) by French members of the Planck satellite team who also study the cosmic microwave background. By the time the information had been taken down from the internet, it had already been widely circulated.

By March, though, the BICEP2 scientists were ready to go public – this was, after all, a big discovery, a significant piece of evidence for an otherwise arbitrarily added part of the big bang model. It was a Nobel Prize contender. Their announcement was highlighted with the typical subtlety of newspaper science reporting by the *New York Times* as 'Space Ripples Reveal Big Bang's Smoking Gun'. Unfortunately, though, the BICEP2 team's triumph did not last long. Within a few months they had to withdraw their findings.

The problem is that gravitational waves are not the only reason that the background radiation could have this particular pattern of polarisation. Polarisation also occurs, for example, when light is reflected – which is why the polarising filters in some sunglasses reduce the glare from reflections in the road. And it was discovered that the cause of the detected polarisation was dust in the depths of space that the cosmic background radiation was passing through, causing a curlicue pattern of polarisation as the microwave photons were reflected by dust particles.

Within weeks of the press conference, American physicist Paul Steinhardt from the Center for Theoretical Physics at Princeton University was writing in *Nature*: 'Serious flaws in the analysis have been revealed that transform the sure detection into no detection.' The Keck array and BICEP3 (both successors to BICEP2 with five times the number of sensors) continue to operate, but at the time of writing have failed to pick up evidence for polarisation from inflation-caused primordial gravitational waves.

Going on, and on …

Inflation, then, is a purely theoretical attempt to patch up a problem with the predictions of the big bang theory that, as yet, has no firm evidence for its existence. Worse, in fact, its impact on the cosmic microwave background radiation should have been detected by now, but hasn't been. One reason for there being support for inflation was that it dealt with one aspect of fine-tuning – the flatness and uniformity of space out of all the possible natures of the universe. But its biggest contribution to the multiverse picture comes with the concept of eternal inflation – ironically first envisaged way back in 1982 by one of the opponents of the multiverse, the same Paul Steinhardt who would later sound the death knell for the BICEP2 results. He felt that it was possible to keep inflation without invoking the anthropic principle, but he did suggest that inflation might not be a limited effect.

At the time, Steinhardt did not see his suggestion as making much of a change to the view of inflation and the multiverse, but he did point out that it was entirely possible that quantum fluctuations – the energy variations that arise inherently from the uncertainty principle – could have invoked inflation in different parts of the early universe as separate events. As he put it: 'I thought I was identifying a feature of the theory – not highlighting a fatal flaw that would eventually come to be known by the community ...'

As is common with this type of theorising, there is no direct evidence at all to suggest that eternal inflation exists. Instead, its supporters, pointing to the relative ease with which new seeds of inflation should have spread (had it existed at all), envisaged that it would give rise to a multiverse of 'bubble universes'. Here inflation runs its rapid course in one region, which then reverts to normal expansion. But it will keep being triggered in different regions, so universe after universe comes into being, of which ours is just one.

The bubble image is quite a useful metaphor. If you imagine a collection of small bubbles, one of which gets inflated, this doesn't stop one of the other bubbles then inflating – they all occupy a wider expanding space, but any one bubble exists without the others encroaching on its internal space. Even so, some theorists have suggested that such a multiverse might be something we could indirectly observe.

If early in the formation of our universe another adjacent bubble came into being, it is just possible that this would have had an impact on the cosmic background radiation as another universe's bubble jarred our adjacent one. This proposition has been investigated for over a decade as the detail of the background radiation has become more precise, but no evidence has yet come to light.

It was, in part, this search that led to the development of the BICEP telescopes (see above) – as it was thought that bubble impacts could also produce primordial gravitational waves (think of the effect of jumping onto a waterbed). These should then have triggered the kind of polarisation that was searched for, but, as we have seen, these have not produced any fruitful results.

Note, however, that even if eternal inflation and the development of our bubble universe is a reality, this has no bearing on the fine-tuning problem, as there is no evidence for or mechanism behind the idea that each bubble universe would have different physical constants from each other … and as we have seen, we can't deduce the existence of multiples universes from fine-tuning. It is arguably more likely that the whole of the multiverse would be fine-tuned in such circumstances.

Such a bubble-based multiverse is not the only possibility to emerge from the esoteric aspects of quantum physics, though. Some physicists believe that the very nature of quantum theory makes it likely that imaginably many universes, each differing by a tiny original starting point from the next, could exist. Welcome to the 'many worlds interpretation'.

MANY WORLDS 8

To date we have been searching for the multiverse by looking outwards, in the region of speculative science inhabited by cosmologists. However, there is another way of producing multiple universes that involves looking inwards – to the quantum world of the very small. If it seems profligate to have a range of universes to deal with the fine-tuning problem, that extravagance pales into insignificance when put alongside the many worlds hypothesis, which deals with a problem in the interpretation of quantum theory by providing extra universes every time there is more than one outcome for the future of a quantum particle.

As we will discover, this can sound somewhat ridiculous – and it is not a view supported by the majority of physicists, though it is held by a vocal minority. The emergence of this apparent requirement for far more universes than we could ever imagine came from a recognised difficulty in dealing with the strange behaviour of energy and matter at the quantum level.

The worlds of Young's slits

Let's start by seeing the origin of the problem that led to the many worlds hypothesis in our understanding of a well-known experiment that is simple enough to be carried out in schools and which dates back to the start of the nineteenth century. It's an experiment that involves a piece of apparatus called the double slits, or Young's slits.

For centuries there have been arguments about the nature of light. Some believed it to be a wave – but were faced with a difficulty in explaining how light could cross the empty vacuum of space. Waves are such a familiar occurrence that it can be easy to miss a simple reality. This is that waves involve the displacement of a medium – *something* has to move in a regular, repeated fashion as the wave travels along. Think of waves crashing onto the beach ... then imagine removing the water but still having the wave. It's not possible. But in space there is no equivalent of water. This is why, to quote the old *Alien* movie tagline, in space no one can hear you scream. We need air to carry the waves of sound.

To solve the lack of a medium, early scientists supporting the wave hypothesis imagined that there was some kind of invisible, intangible substance known as the 'luminiferous aether' that filled all of space and through which the light waves travelled. The problem with this apparent solution was that there wasn't anything obvious there. What's more, for light to be able to travel vast distances from the stars, this material would have to be incredibly rigid, as the floppier a substance is, the more the energy of a wave gets dissipated, leading to it fading away and disappearing. Light appears to be able to travel forever.

Others, of whom historically the most notable was probably Isaac Newton, believed that light was made up of tiny particles, originally called corpuscles. If this were true it would be easy for light to travel across empty space, as there is no problem with a spray of particles moving through a vacuum. The approach also worked well with some of the properties of light such as reflection, where we can imagine particles bouncing off a surface like a ball. But the particle theory provided difficulties when it came to some of the more arcane behaviour of light, such as refraction: the change of direction that occurs when it moves from substance to substance.

The English polymath scientist Thomas Young seemed to hammer an irresistible nail into the coffin of corpuscles with an experiment that he carried out in 1801. Young sent a beam of light through a narrow slit, after which it spread out for a while before hitting another barrier with two parallel slits in it. A little distance further on from this was a screen. The neat part about this simple experiment is that it provided a situation where particles should behave differently from waves, making it possible to determine light's nature.

If a beam of light were composed of many particles, we would imagine that some of those particles would go through each of the two parallel slits, producing two corresponding blobs of light when the particles reached the screen. But if light consisted of waves, things would be different.

Interfering waves

Imagine two sets of waves, one setting out from each slit. These are so-called transverse waves, like ocean waves, where the waving is at right angles to the direction of travel, involving, say, moving up and down as the wave travels along. With the right distance involved, before those sets of waves reached the screen they would pass through each other. There was a well-known effect called interference that occurred particularly when two waves of the same wavelength* met.

If both of the waves were moving up at the same point, then the result would be reinforcement – the waves would combine to move the medium upwards more. If one wave was moving up while the other was moving down, they would cancel each other out, producing a point in the material that is hardly moving at all. This can be clearly seen when dropping two

---

* Wavelength is the distance travelled by the beam of light between, say, the highest point in the wave's up and down motion and reaching that same height again.

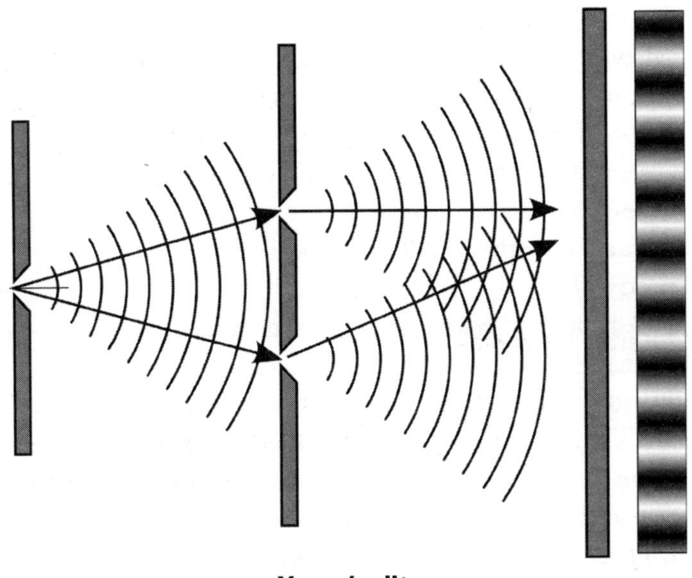

Young's slits

stones into a still pool of water and watching the waves interact.

If this kind of interference happened between the light travelling from the two slits in Young's experiment, you would expect to see bands of light on the screen where the waves were reinforcing each other, and bands of darkness where they were cancelling each other out. And that is exactly what was observed.

The assumption at the time Young undertook this experiment was that he had definitively proved that light was a wave. To an extent this was problematic, as many experiments failed to discover any evidence that there was an aether for the waves to travel through. But the idea of light as a wave would be reinforced by the realisation of Scottish physicist James Clerk Maxwell that light was an interaction between magnetic and electrical waves. A moving electrical wave generates magnetism, while a moving magnetic wave generates electricity. If these waves travel at exactly the right speed, they will sustain each other – and that speed turned out to be the speed of light.

Although Maxwell himself never accepted it, this theoretical explanation of the nature of light even did away with the need for the aether, as electrical and magnetic waves do not require a medium to travel through.* But the apparent definitive solution was short-lived because discoveries in early quantum theory at the start of the twentieth century would show that light also must be a stream of particles, later known as photons. This was because it exhibited a behaviour that was just as definitively that of particles as interference appeared to be of waves.

Light as particles

The realisation of the particle-like behaviour of light was the theory that won Albert Einstein the Nobel

---

* The great English scientist Michael Faraday developed the idea of fields as a kind of non-substantial medium that filled all space. In this picture it is the field that is varying in a wave-like manner. Field theory would become central to the development of twentieth-century physics.

Prize, rather than his now better-remembered work on relativity. This was based on the photoelectric effect, where light shone onto a material would generate an electric current. It had been found that more energetic light (towards the blue end of the spectrum or higher) would produce electricity, but as the energy was reduced (heading towards red and beyond) the production would suddenly stop.

If light were a wave, you would expect that lower energy might reduce the amount of electricity produced, but not stop doing so at a specific cut-off point. However, if light were a stream of particles, each individual particle would have to have enough energy to knock an electron out of the material to form part of the electrical current. Drop below a certain energy level for the incoming particles, and no current would flow. Einstein's theory predicted exactly what was observed. But now there seemed to be definitive evidence that light was both a wave and particle.

The apparent paradox that light somehow had this dual nature (and the same duality was discovered to apply to tiny matter particles such as electrons) would be solved by a realisation emerging as quantum physics was developed that the fundamental wave here was not physical in the conventional sense, but one of probability. It seemed that until they interacted with something else, quantum particles did not have definitive locations. After passing through the twin slits in the experiment, all that existed was a range of probabilities for photons to be in any particular location.

This approach worked well in terms of accurately describing what appeared to happen. If light were indeed a stream of photons, then the calculations of the probabilities of a particle turning up at a particular point on the screen produced exactly what was observed – a series of light and dark fringes as if it were a wave, but as a result of particles obeying these probability waves.

Perhaps most remarkably of all, as physicists' ability to control light became better it proved possible for photons to be sent through the apparatus one at a time and still the pattern of light and dark fringes emerges over time. The same applied to other quantum particles. This is sometimes described as each photon travelling through both slits at once and interfering with itself, but this isn't a good description. In reality, according to the most widely accepted interpretation of quantum theory, the probability wave describing a photon's location includes a passage through each slit – the particle is nowhere specifically, but has a probability of being in very many places until there is an interaction with the screen.

Quantum interpretations

However, there is a built-in proviso to that 'most widely accepted interpretation' description. From the earliest development of this 'Copenhagen interpretation'* in the

---

* The Copenhagen interpretation is named after the Danish base of leading quantum physicist Niels Bohr, rather than a particular meeting at which the concept was thrashed out.

1920s, some physicists have been uncomfortable with the idea that physical processes do not have 'reality' until there is an interaction causing a so-called collapse of the wave function – the mathematical probability calculation using Schrödinger's equation that specifies where the particle might be.

This equation works wonderfully until a measurement is made, when the 'collapse' occurs and suddenly the particle has a defined location with no mechanism provided to explain how this transition occurs. Although justifications have been developed, it remains a disputed situation. The outcome has been a plethora of alternative quantum interpretations that attempt to do away with the wave function collapse. Frustratingly, though, in most cases there is no test that can be made to distinguish between the interpretations. Each predicts the same outcome. The interpretations don't affect quantum theory's incredibly accurate prediction of what is observed, which has enabled the development of all modern electronics. But they do worry those who want a good philosophical basis for their science. (The alternative approach to not worrying is sometimes disparagingly called 'shut up and calculate'.)

This isn't the place to go through all the different possibilities for an effective interpretation (see the Further Reading section for a deep dive into this), but one of the better-supported alternatives brings us into the realm of multiverses. This, as we have seen, is known as the many worlds interpretation, which at first sight involves a dire contradiction of Occam's razor. The razor is a philosophical approach to deciding which theory is better when there is no direct evidence to help choose – with no other reason for selecting it, go for the approach that involves least complication. As William of Ockham, the fourteenth-century English philosopher behind the razor, put it, 'Entities are not multiplied more than is necessary.'

When we examine the outcome of a quantum interaction, a very effective mechanism devised by American physicist Richard Feynman to describe and calculate the outcomes is known as 'sum over paths'. For example, in the classic 'twin slits' quantum experiment mentioned earlier, all traditional quantum theory tells us is that there is a distribution of probabilities of taking any possible path en route from the source to the final possible destinations on the screen. Most of these paths cancel out when the sum is taken, but there are still many potential outcomes. We can't predict where a specific photon will end up.

Those who support the many worlds hypothesis consider each possible outcome to exist in a different parallel universe. Given that every quantum particle in existence has the potential to have many possibilities for its future, this makes for an unimaginable number of parallel universes in the many worlds multiverse – which surely feels like entities being multiplied far more than would have pleased William of Ockham.

The many worlds multiverse

This potential multiplicity of worlds began in the mind of a Princeton graduate student called Hugh Everett III in 1957. At the time, the Copenhagen interpretation of quantum theory was hardly ever questioned, but Everett was not comfortable with its implications – specifically the need for the collapse of multiple possibilities into a single observed outcome.

For a graduate student to challenge the worldview of the majority of physicists took a particular kind of personality. I have seen Everett described as both highly intelligent and cold in his attitude to humanity – in effect, the classic mad scientist. We hear that he was cynical, distrustful of authority and presented a superior attitude to the world. Part of this, specifically the distrust of authority, isn't a bad thing for scientists. It is necessary to disregard authority to make breakthroughs in science. It's not for nothing that the Royal Society's motto is 'Nullius in verba', which roughly means 'take nobody's word for it'. In effect, go with the evidence, not what an expert tells you. But it's a difficult attitude to maintain when, to some degree, your career depends on those authorities. It is only sustainable when combined with a degree of genius, which some believe Everett had.

It has been suggested that Everett might have been inspired to think of many parallel universes by his time at the Washington DC based Catholic University of America, where he gained his first degree (in chemical engineering). At the university he was required to take courses that strayed into theology and philosophy, and he is likely to have come across the German mathematician and philosopher Gottfried Leibniz's suggestion that God was able to consider every possible reality to come up with 'the best of all possible worlds'. It might only have been in God's head, according to Leibniz, but that description gave a picture of a kind of multiverse featuring branching alternatives.

It's an interesting thought, but such random inspiration feels a little too much like the crutch often used in second-rate mystery stories, where the detective is inspired to think of a solution to a crime by experiencing something that reminds them of a possibility or inspires a new line of thinking. It seems unlikely that Everett, driven by the need to avoid the condensing down of the multiple worlds implicit in Richard Feynman's quantum approach to a single specific outcome, would need such an inspiration.

Everett's doctoral thesis, originally titled *The Theory of the Universal Wavefunction*, referring to there being a single underlying mathematical progression of all possibilities, but renamed the less provocative *On the Foundations of Quantum Mechanics*, was his one shot at the field. Rather than continue to fight the establishment, he would soon move into a safer area of work (ironically, as it's not usually associated with safety, this was defence). But the baton of the many worlds interpretation would be taken up by another American physicist who was already well-established as a professor, so less dependent on the opinion of others: Bryce DeWitt.

DeWitt's speciality was gravity – in the late 1950s there was a revived interest in the kind of unification of gravity and quantum physics that Einstein had spent most of his work life on after developing the general theory of relativity, with no useful outcome. At a conference DeWitt organised, he was given a copy of a paper by Everett on his many worlds hypothesis. DeWitt was fascinated but doubtful – specifically because of the speculative nature of the interpretation, which seemed unable to be tested. But on hearing that DeWitt criticised many worlds by pointing out that he clearly experienced his own person as a single entity that did not branch into many universes, Everett's final sally was to note that those who denied the movement of the Earth implied by Copernican theory would similarly point out that you can't feel the Earth moving.

This seems to have pushed DeWitt into giving the many worlds approach greater consideration. It was his 1970 article on the quantum interpretation in the widely read *Physics Today* that brought it to a wide audience – and a growth in support that looked for a while as if it would make it the dominant interpretation, though in practice it has been of more interest in popular-science books and magazines than it has been useful to working physicists.

A blow to many worlds

Although many worlds captured widespread attention in the 1970s and 80s, other new interpretations have since become more fashionable (though the majority of physicists continue to stick with the straightforward approach of not particularly worrying about interpretations and just getting on with the science). However, on 6 January 2025, the magazine *New Scientist* carried the headline 'The End of the Multiverse?'

It has to be acknowledged that for some years *New Scientist* has had a tendency to run with speculative clickbait-style headlines, and the question mark at the end of that statement remains suggestive of this – a true discovery would warrant an exclamation mark instead. But overall, the concept presented had some merit. The story was not about the potential existence of multiverses as a whole, but rather the many worlds interpretation in particular. University of Bristol physicists Sandu Popescu and Daniel Collins claimed to have 'essentially demolished one of the arguments for [many worlds]'.

As we have seen, the many worlds interpretation was introduced to cope with the oddities arising from the probabilistic nature of quantum physics. This unspecific nature for the outcome is fine, for example, when considering location. But some measurements are subject to conservation laws. If we consider properties of a particle like its momentum or angular momentum, they can only change if a quantity is transferred elsewhere because, like energy, these properties are conserved. This means that the total value in a closed system where nothing goes in or out cannot change. If it's also the case that all that exists for one of these quantities is a collection of probabilities in what is described as a superposition of states, until there is an interaction with the particle's surroundings, what happens to conservation laws, which many regard as central to the working of physics?

This is, to say the least, a matter of dispute. Fans of the many worlds approach would say that this is one of the clear reasons to support their interpretation. If you add together the momentum, say, of all the particles in the vast range of universes, the result will be to conserve momentum overall, even if it isn't the case in any specific universe. Those who stick with the conventional approach developed from Bohr's Copenhagen interpretation would simply say there is nothing to worry about, because all you are dealing with in the superposition is probabilities, not actual momentum being gained or lost.

What Collins and Popescu did was to consider the combination of the mechanism used to get the particle into a superposition of states and the particle itself. One of the most fundamental aspects of quantum physics is the concept of entanglement. This is a phenomenon where two quantum systems – which can be individual particles or large accumulations of particles – become connected in a way that a change in one system is instantly reflected in another. In effect, the entangled entities become part of the *same* system even though they can be widely spatially separated. Such entanglement has now been demonstrated across a distance of around 2,500 kilometres (1,600 miles) via a Chinese satellite called Micius.

The Bristol pair discovered that in specific examples of angular momentum, entanglement between the particle being studied and the mechanism getting it into superposition meant that whatever the final measured value of the particle's angular momentum was, the angular momentum of the mechanism would reflect this, such that their overall combined angular momentum was conserved.

This meant that the argument supporting many worlds from the apparent lack of conservation of angular momentum no longer applies – and it may also be the case for other conservation laws. And this is the reason behind that highly exaggerated headline. As yet, this outcome has not been determined experimentally, but the theoretical basis for dismissing a problem with conservation has not been challenged.

Supporters of the many worlds interpretation would point out that the conservation laws were not by any means the sole reason for supporting their interpretation of quantum physics. The primary reason was to do away with the need for the collapse of the superposition, with has no obvious physical mechanism to explain it. Somehow, a set of probabilities becomes an actual measurement. But it's certainly true that it is one less argument in support of this particular take on the multiverse.

As always with a speculative theory like this, unless it's possible to devise some kind of test that would distinguish the 'many worlds' multiverse from the more intuitive single universe, it must remain something that sits on the very borders of science and cannot truly be considered a scientific theory. But one distinguished supporter of many worlds has suggested there is (in principle) a way to test for it. This is the British physicist, David Deutsch.

The self-aware quantum computer

Deutsch's speciality is quantum computing – he was the first to suggest that a computer that made use of the remarkable behaviour of quantum particles, rather than using the simple 0 or 1 capacity of a traditional computing bit, would be able to undertake computations that could never be duplicated by a conventional machine in any practical timescale. He has suggested that it would be possible to detect the existence of the many worlds multiverse by making use of a self-aware quantum computer.

In recent years, artificial intelligence (AI) has become a central theme in the development of computing, but Deutsch had something different in mind from the capabilities of a generative AI such as ChatGPT, Gemini or Copilot. His self-aware concept was a quantum computer that was able to monitor its own state. Clearly in a conventional view of there being a single universe, such a computer would only need to handle the single state that it can be observed to be in at any one time. But in a many worlds multiverse, something different would happen.

According to Deutsch, 'what [the computer] tries to observe is an interference phenomenon between different states of his [sic] own brain. In other words, he tries to observe the effect of different internal states of his brain in different universes interacting with one another.' The power of a quantum computer is that its equivalent of bits, called qubits, which are quantum particles, can interact with each other, expanding their ability to undertake computation. Deutsch envisaged this kind of interaction occurring across the different worlds of the multiverse.

Few think that Deutsch's vision is a practical tool for seeing if many worlds truly exists, but some have suggested a much darker way that the existence of this kind of multiverse can be checked out. All it takes is a gun and a remarkably strong faith in the many worlds interpretation, where the experimenter is prepared to put their own life at risk.

The test involves the scientist shooting themselves in the head.* If you think of all the possible outcomes from that moment of pulling the trigger, there will be a few where you don't get hurt. For example, the gun might jam or the cartridge could be a dud. In the majority of universes, you will indeed be dead. But, much like the anthropic principle, we can argue that if all the outcomes do indeed exist, then there will be universes in which you survive, and inevitably you will only be conscious of those locations. As a result you would not experience death. And you can repeat this as often as you like.

I include this not as a suggestion for a realistic test anyone could make, but to show how weak an argument this is. It certainly couldn't be seen as a scientific test. After all, it only provides evidence to the conscious you – in the vast majority of cases the rest of the world would see nothing more than a suicide and there would be no evidence provided to support the theory.

---

* It really shouldn't be necessary to say this, but please do not try this extreme test of the many worlds interpretation out at home (or anywhere else).

There is something medieval-feeling about such a test making use of drastic means. It feels not dissimilar to attempting to identify witches by throwing them in water and seeing if they sink or float. It is perhaps time to head back to what feels like sounder science if we consider a way that multiple universes might exist and interact thanks to an implication of another branch of speculative science: string theory.

COLLIDING UNIVERSES 9

Although it is entirely possible for a multiverse to exist without there ever being any interaction between the universes within it, one possible explanation for the big bang in our universe is that it is the result of collision with a neighbour – a possibility that originates in string theory (or, more precisely, its extension M-theory). This provides an attempt to give a unified structure to the zoo of fundamental particles, and to combine the different forces of nature. To see how this concept comes into being, we first need to get a handle on strings.

What if particles were strings?

Here, as with many worlds, we are dealing with what initially appears to be the total opposite of cosmology and the big bang – instead of taking in the whole of the universe, we are zooming in on the very small, uncovering the fundamentals of physics. Most forces and particles are very effectively described by quantum physics, but gravity, underpinned by Einstein's general theory of relativity, is not a quantum theory. Despite nearly a century of effort to combine relativity and quantum physics, they remain incompatible.

Yet, although quantum physics is generally thought of as applying to very tiny parts of reality, it has a part to play in cosmology when we think of the origins of the universe, effectively beginning as a dimensionless point immediately before the big bang. This means that the incompatibility between quantum theory and gravitation – the force at the heart of cosmology – is at best embarrassing. For a long time, one of the leading attempts to deal with this intellectual conflict has been string theory.

Following on from Newton's corpuscles, we usually think of the building blocks of matter – and the mechanism for transferring forces from place to place, known as force carriers – as particles. These are point-like objects that are very small or, in some cases, dimensionless. The standard model of particle physics includes seventeen such particles from the relatively familiar, such as the electron, to the most recent addition, the Higgs boson. But physicists like things to be as simple as possible – and string theory does away with this collection of different particles, replacing them all with a single concept called a string.

Unlike a particle, strings are not point-like (though they are still extremely small) – they can form loops and can vibrate in different ways, producing families of 'particles' in the different ways they vibrate. It ought to be stressed that these aren't literally string-shaped objects – apart from anything else, they are one-dimensional, having no width. They are theoretical constructs that have string-like behaviour, just as, for example, light isn't really a wave or a particle, but a natural phenomenon that has both wave-like and particle-like behaviour.

In one sense, then, string theory provides a very useful simplification of the seventeen different particles. But just as the many worlds interpretation does away with an oddity of quantum physics by introducing a different sort of complexity, so string theory makes some significant demands on physicists who want to make use of it. Specifically, string theory requires not the usual four dimensions we have met regularly so far (three of space and one of time), nor even the five of Kaluza and Klein (see page 52), but ten dimensions. And before we get to colliding universes, we will need to move on to string theory's big brother, M-theory – which requires yet another dimension.

The missing dimensions

When we first looked at the early ideas of parallel universes and existing in a 'different dimension', what was being thought of was in effect a separate set of three or four dimensions from the ones we directly experience, working in exactly the same way as the familiar ones. Kaluza and Klein then added another dimension to existing spacetime to try to combine gravity and electromagnetism, but could not make the model match reality.

This takes things even further – these are six or seven dimensions extra that would form the structure of our own universe, not something entirely separate, required to enable the appropriate vibrations. And that being the case, an obvious question is where all these extra dimensions are. Why don't we experience them?

As was the case with Klein's single extra dimension, according to the mathematics that requires their existence in string theory, these extra dimensions are curled up, not unlike an incredibly thin drinking straw. This is the favoured analogy of many a string theorist trying to describe these packed-away dimensions. They might say, 'If you imagine a really, really tall drinking straw, then from a distance it will look like a one-dimensional line, but it really is three dimensional – only you can't detect the other two dimensions.'* While this is approximately true, if you can see it at all, it can't only have one dimension or it would have no thickness – so perhaps it's not the best analogy.

One difficulty with the 'curled up' dimension is that the familiar spatial dimensions are effectively infinitely long lines. So, we are being asked to envisage an infinitely long line curled up so tightly that it has no measurable dimensions. Mathematically this is feasible as the 'line'

---

* Often physicists might say a drinking straw is more like two dimensions – one of the length and the other curved around, but that's only true in our 3D world if the straw has been slit along its length and unwrapped, where the whole point of string theory's hidden dimensions is that they aren't unwrapped.

of the dimension has no width, but it doesn't make for an easy mental picture.

However you imagine it working, these curled-up dimensions are considered to be so small that we can't detect them. Note that there is no experimental or observational evidence for their existence – they are merely necessary to make the maths work. It's true that the theory has the potential to fit our universe, but unfortunately the reality of our universe is just one of many possible solutions. There are more possible solutions of string theory than there are believed to be protons in the universe – with no reason for selecting our reality. Theories that don't make testable predictions are arguably of limited use.

A huge amount of effort has been put into string theory and M-theory ever since the first ideas began to emerge at the end of the 1960s. There are theoretical physicists who have spent their entire careers on the theory without ever seeing any evidence to back up the mathematics they have so lovingly constructed. Some suggest that a reluctance to change direction after investing so much time and effort is the sole reason the theory is still given significant attention. Its detractors have described it as 'not even wrong'.*

---

* 'Not even wrong' is a favourite insult in physics, first coined by the Austrian physicist Wolfgang Pauli. It's not entirely clear what he first applied this to. He may have come up with it in the 1940s as a response to one of the stranger possibilities in quantum physics, but he certainly then applied it to the many worlds interpretation. Since then, it has been used by critics of string theory, most notably by American physicist Peter Woit, who wrote a book of this name on the subject.

Magic, mystic or membrane

From the viewpoint of looking for a possible multiverse, M-theory offers more hope than basic string theory, though it starts from a similar place. There are five major variations of string theory, which can be combined by adding an extra dimension, producing M-theory. Exactly what the 'M' in the name is has never been definitively established. Its originator, physicist Edward Witten, has suggested it might be magic, mystic or membrane … a vagueness that could also make it something of a mockery.

One particular problem with any M-theory approach is that it provides around 10^{500} (1 with 500 zeroes after it) possible solutions, all of which the most extreme multiverse enthusiasts could describe as fitting a different universe in its own right. If you are prepared to go for speculation squared and combine M-theory with eternal inflation (see page 95), some physicists suggest the two will result in that number of universes, each with a different cosmological constant. But that's a whole lot of supposition. Leaving aside the fact that there seems to be no particular mechanism, given a lack of communication between universes, of allocating each solution to a different universe, this leaves us with a theory that has no apparent connection to any particular universe (ours).

Given this lack of clarity, a number of leading physicists, notably the late Steve Weinberg, an American physicist who won a Nobel Prize for his work on combining in theory two of the fundamental forces of nature (electromagnetic and the weak interaction), re-introduced the good old anthropic principle, arguing that while there are many possible universes, only a relatively small number could refer to ours, as we're here to observe it, and most couldn't support life. Similarly, only a small percentage would happen to have a particularly small cosmological constant (which is true of our own universe), while the rest would have a very large one. So, they argue, there aren't many solutions we need to consider.

Although this does feel a bit of a fudge, it has some similarities to a process known as renormalisation, which proved necessary to sort out issues with quantum electrodynamics, the part of quantum physics describing the interaction of electrically charged particles and photons of light. The initial calculations produced some infinite values – but this was effectively patched up by replacing calculated values with those measured from experiment. This worked dramatically well.

However, renormalisation made calculations work that up to that point were impossible to carry out – resulting in a major contribution, for example, to the development of electronics. By contrast, using an anthropic filter to select a relatively small number from the ridiculously vast number of possible outcomes from M-theory does not add any useful insights. Where renormalised quantum theory makes plenty of predictions (leading to its use in electronics), M-theory constrained by the anthropic principle still makes no predictions.

Paul Steinhardt, who was so scathing about the eternal inflation that he had accidentally conceived, is equally dubious about string theory and M-theory. He has commented: '[String theory] doesn't produce a single type of particle physics. It produces a nearly infinite variety, like the multiverse, an exponential number of different possibilities called a landscape of possibilities. And if that landscape picture is correct, it actually has the same sort of problem that inflation does, which is that it gives too many possible outcomes. And so, like inflation, it's not really telling us why things are one way versus another way.'

The brane-based multiverse

Although Steinhardt is highly sceptical of the relevance of string/M-theory here, he has still theorised about a possible multiverse that is based on an implication of M-theory. Working with South African physicist Neil Turok and American physicist Burt Ovrut, he devised a hypothetical multiverse known as the ekpyrotic universe. This Greek name, suggesting things started from fire, was not so much the result of a search for a multiverse as an explanation for how the big bang was initiated. The problem with our best accepted picture of the origins of the universe is that it's not at all clear how the initial conditions immediately before the big bang came about. The ekpyrotic universe explains this.

To get to this picture, we need to know that in M-theory, those extra dimensions don't have to be rolled up because we are no longer limited to one-dimensional strings – there can be membranes (usually shortened to branes) of multiple dimensions. This gives us a potential picture of a multiverse taking up the ten spatial dimensions of M-space where the universe we inhabit is just a brane made up of three spatial dimensions and one of time, floating in the bigger reality. And there is no reason, if this were the case, why there should only be one such brane.

The trio imagined a pair of branes floating in that M-space multiverse. These could be former universes that have reached heat death. This is one possible end to the universe. The second law of thermodynamics, one of the most important laws of physics, tells us that in a closed system (like the universe) entropy – the measure of disorder in the system – increases.

It might seem that this doesn't reflect reality. The 'order' described by entropy shows us how many ways the components of a system can be arranged. For example, if you think of the letters, numbers and symbols that make up the text of this book, there are far fewer ways to have them organised as they currently are than there are to have a complete jumble. So those characters have lower entropy here than they would if they were separated from the page and shaken up in a bag. We can see how entropy tends to increase rather than go the other way by thinking of dropping an egg on the ground and watching it shatter. This is a far more likely process than all the mess on the floor reassembling to make a complete egg.

Of course, it is perfectly possible to reduce entropy. The atoms that went into the egg were far more disordered before the egg was formed. But the trick with entropy is that if you put energy into a system, you can reduce entropy. This is why complex structures and organisms have emerged on Earth despite running opposite to the increase in entropy – because the Earth is constantly gaining energy from the Sun.

If we think of the end of the universe, though, eventually there will be no elements left to go through atomic fusion in stars producing energy. The stars will go out. Without sources of energy, over time entropy will prevail. Everything will be brought to an end in unstructured nothingness with no significant energy to make anything happen. But in an ekpyrotic multiverse, this does not have to be the end of the story. According to Steinhardt and his collaborators, some branes would be attracted to each other until they collided.

Initially, this seems an unlikely premise. After all, these branes are not within the same universe – each forms a universe in its own right. And as we have seen, we would not usually expect any interaction between separate universes – it's almost the definition of them being a universe in their own right, rather than a region of our own. But the possibility that they would be attracted to each other is suggested by something strange about gravity. It's very weak.

Leaking gravity

This might seem a bizarre suggestion. Try falling off the top of a mountain (or, slightly less dangerously, picking up a car) and see if gravity feels weak. Yet, compared with the other forces of nature, it is incredibly feeble. This is demonstrated by a scientific experiment most of us can do in the home, using only a fridge magnet and a fridge. Take a magnet off the fridge and hold it very close to the metal without quite touching. Let go. There are two fundamental opposing forces working on the magnet at this point. One is the gravitational pull of the Earth, comparatively massive to anything else nearby (around 6 billion trillion tonnes). The other is the electromagnetic force produced by the tiny magnet. The magnet wins. It's difficult to make a direct comparison, but, roughly, electromagnetism is 10^{36} times stronger than gravity.

One suggestion as to why gravity is so weak is that the gravitational force leaks out of our universe's brane into the wider multiverse. And this would enable two branes to be drawn towards each other. When such monstrous objects collide, the result is a vast dispersal of energy, starting off new expansion in a universe that had ended up without internal structure – from inside our brane this looks very like the big bang.

Because our universe would continuously expand in such a picture, although its overall entropy would be high, that entropy would be very thinly spread – the result would be just the kind of conditions that are envisaged as being necessary at the start of a universe like ours, where the collision with the other brane would then inject the oomph of energy to get things started all over again.

If this picture were true, the result could be a cycle in which a pair of branes repeatedly bounced away from each other. The distance between them would expand, but gradually their speed of separation would slow due to gravitational attraction, until there was a turning point and the branes started to accelerate towards each other again, until the next collision occurred. This picture neatly does away with a single moment of creation for our universe. The multiverse as a whole could have been undergoing such internal collisions forever, doing away with a starting point for time – although there is nothing to prevent such a multiverse having its own moment of creation.

This helpfully solved a problem of the kind of cyclic universe that had previously been envisaged. Broadly there are two types of cosmological history for a universe – having a moment of creation or running endlessly through a series of cycles. Some cosmologists prefer such continuing cycles because the scientists are concerned that having a creation has too strong religious connotations. But there is a thermodynamic problem with a universe that repeatedly expands and contracts with intermittent bangs – it should run out of energy over time, undergoing an entropic progression that echoes the heat death of the single universe, but happening over a far longer period of time.

From steady state to cyclic

Before the big bang theory became dominant, the leading cosmological theory was the steady state theory, developed by Fred Hoyle, Thomas Gold and Hermann Bondi.*

---

* The scientists behind the steady state theory claimed it was inspired by going to see the film *Dead of Night* at the cinema in Cambridge. This brilliant 1945 compendium horror film begins and ends with exactly the same scene – so effectively has no beginning.

This had no beginning of the universe, just a constant expansion with creation of matter constantly happening to avoid everything thinning out to nothing. Unlike early cyclic universes, this didn't have an entropy problem (though it was necessary to envisage a new physical field able to bring matter into being).

Unfortunately for the steady state theory's supporters, as telescopes became able to look further into space, and hence further back in time due to light taking time to get to us, it became obvious that the early universe appears very different from the current one. The ekpyrotic universe provides the same benefit as steady state – expansion is the only change, and so it gets around the thermodynamic issues of expansion, contraction and banging – without the need for the past to look like the present.

In chapter seven, we met the tricky concept of inflation – yet another cosmological addition to theory with no evidence yet to back it up. As we have seen, one reason that inflationary theory was originally devised was to explain how the universe had expanded so quickly in its formative fraction of a second, enabling it to be as uniform as the cosmic microwave background radiation suggests it was. But the brane universe starts off almost uniform just before the branes collide – there is no need to have a ridiculously fast period of expansion to explain it.

The ekpyrotic picture has another potential benefit. It allows theorists to explain away what is generally considered the biggest difference ever discovered between a theoretical prediction and an actual measurement.

This is the scale of lambda, the cosmological constant behind dark energy – the theoretical reason the universe is expanding at an accelerating rate. There is a quantum effect that is usually considered to be behind dark energy called vacuum energy, which we encountered in chapter seven when considering the 'false vacuum' that was envisaged to support inflation. Unfortunately, the cosmological constant is far smaller than theory predicts – it's a factor of 10^{120} out (that's 1 with 120 zeroes after it). That is unbelievably wrong.

However, if the universe were vastly older than the around 13.8 billion years stretching back to the big bang, then it is feasible that the value of the constant could have fallen to such a low level. And if the ekpyrotic universe were the real deal, it could have been around for so long that the cosmological constant had decayed to its present value.

Bearing in mind Steinhardt's dislike of multiverses, this approach (renamed the cyclic universe in a later incarnation, probably because no one could remember how to spell ekpyrotic) was a brilliant example of having your cake and eating it. Strictly speaking it *does* involve a multiverse, as it requires another brane universe to collide with our own. But it doesn't have any bubbles invoking anthropic arguments, or multiple alternatives to deal with fine-tuning. In this picture, we exist in a single universe with no beginning and no end in time, which has potentially gone through many stretching cycles. The theory doesn't need to worry about what's happening on other branes (or even if there are more than two).

Brane collisions versus inflation

This 'brane world' approach also does away with any need for Steinhardt's bugbear of inflation – the period of time before the collision involves a smoothing out of the universe and does away with the need for the strange starting and stopping phenomenon of inflation. This is wonderful indeed, and gives us a reason to have a multiverse without the poor logic involved in the attempt to explain away fine-tuning.

Not only that, but there is a clear, testable distinction between a brane world collision and inflation. As we have seen in chapter seven, the aftermath of inflation should be seen in the cosmic microwave background radiation in the form of polarisation from gravitational waves, while the brane collision would not produce strong enough waves to have a detectable effect. If we could detect the aftermath of these primordial gravitational waves, it would rule out a brane collision, while the current lack of evidence with more and more sensitive studies makes inflation unlikely (though, of course, it is not positive evidence for branes).

Going with some kind of cyclic universe with branes does, however, require us to take on M-theory (or something like it) with all the weird and wonderful implications of ten spatial dimensions – without any way of checking that theory by using actual observations and experiments.

It's also worth saying that we have no reason for assuming that the time taken for a cycle would be sufficient for the universe to have reached the wide distribution and low density of entropy necessary for what the cosmic microwave background tells us about the beginnings of the universe. As a result, we are effectively adding yet another fine-tuning component to our existence.

As a cosmology based on M-theory feels like mathematics looking for a real world to describe, it can seem that giving any credence to an M-theory-based multiverse is building supposition on speculation without any value other than intellectual entertainment. It's true that a cyclic universe disposes of some of the issues in other multiverse styles – but it is as yet another theory without any useful way to test its validity.

This leads us neatly on to what is arguably the most speculative of all multiverse ideas – that we live in a hologram. This might sound like a more visual version of the virtual multiverse we visited in chapter three. But here it would reflect an odd but true physical reality.

HOLOGRAPHIC UNIVERSES 10

Our starting point for a dive into the existence of holographic universes is to revisit the black hole.

Black holes and holograms

Black holes are now a familiar concept from science fiction, and are even used as a metaphor for a gap in budgets. At the time of writing, the UK government repeatedly complains that its predecessor left it with a '22-billion-pound black hole in the public finances' to the extent that *The Times* newspaper's political podcast has run a sweepstake on how early the comment would appear in parliament at Prime Minister's Questions. But (unlike multiverses) there is good evidence that something like black holes do exist.

As we saw in the opening chapter, these are stars that have undergone catastrophic collapse, so much so that all the former matter making up the star ends up compacted as a dimensionless point. The reality is that we can't closely examine a black hole, and are never likely to know for sure exactly what is happening behind its event horizon, from which no information can escape. If black holes truly were dimensionless, our current physics would break down as its gravitational pull went infinite – so it may be that what we identify as black holes are something close to the concept but which didn't quite make it to total collapse due to some unknown restraint. However, there are definitely astronomical bodies out there that bear a close resemblance to this prediction from Einstein's general theory of relativity.

To get to a holographic universe we also, of course, need to be familiar with holograms. While a hologram stores information in a two-dimensional medium, unlike a photo or a movie it manages to compress the information required to see three-dimensionally into that flat source. If you think of the difference between looking at a photograph of a view and the actual view seen through a window of the same size, the difference is that with the window, as you move what you see changes. Because the light is coming through the window in different directions, the resultant image you experience is dependent on your position.*

---

* What a hologram can't do, despite what's seen in plenty of science fiction, is project a 3D image, moving or still, into empty space. It always involves either looking through the hologram, as if through a window, or looking at a screen which it is projected onto, though these screens can be made relatively hard to see.

This is achieved in a holographic image by encoding extra information that is related to the direction taken by the incoming photons. To get to the potential existence of holographic universes, some physicists suggest that our perception of three spatial dimensions is not a reflection of reality, but rather the experience of existing in a curved, two-dimensional holographic space around a black hole. This might seem an unnecessary complication, adding something that we can't be aware of to the structure of reality, yet the mathematics behind the concept do appear to work, even if they can't distinguish whether or not it is the case.

It from bit

There's one more component we need to see how this could even be possible, and that's the link between the universe and information, something memorably labelled by American physicist John Wheeler as 'it from bit'. This considers all existence to be nothing more than streams of data. Not a simulation, as seen in the virtual universes of chapter three, but rather providing the fundamental nature of reality. This is not information residing in a computer: the universe *is* the computer as a result of its information component.

It can be easy to oversimplify with a concept like this. 'It from bit' supporter Seth Lloyd, an American physicist, has said, 'The universe is made from bits. Every molecule, atom and elementary particle registers bits of information.' But this seems to confuse the distinction between something capable of registering information and being made of it. Think of a book, lying on a table. In one sense that book contains many bits* of information. But there is a simpler registration that comes from the orientation of the book, holding a single bit.

The book can be face up or face down – a single bit with two possible values. It might not have been intentionally stored information, but the information is nevertheless there. If we shared knowledge of the meaning of that bit, I could use it to pass a message to someone else, depending on how I placed the book on the table. So the book registers a bit of information from the way it was put down – but we can't say the book is made from that bit.

Taken as a broad viewpoint of the universe as a whole, though, we could see the vast interacting system of quantum particles or fields as being a constant exchange of information, embodied within the universe, and in effect that information flow is sufficient to represent the universe in total – this seems to be closer to Wheeler's vision. The information doesn't displace the reality of matter and energy, but can be seen as a more fundamental description of the same thing.

Before we get to how black holes come into this it's worth taking a brief dip into what you might see as a more extreme version of 'it from bit' – that mathematics not only can be used to describe the universe, but is its essential nature. Such a mathematical universe gives us

---

* For our purposes, a 'bit of information' is being used in the computing sense of a single binary value, 0 or 1.

the potential for a variant multiverse. This idea is not mainstream, but is championed by Swedish-American physicist Max Tegmark.

A multiverse made of maths

The history of science and particularly physics and cosmology has shown an increasing dependence on maths – its formulae and numbers enable us to model the universe, describe what is happening and make measurements from experiment and observation. But Tegmark thinks we should go further. He argues that our descriptions of what's happening – the words we use – are human in origin. But the underlying mathematics is more universal and would be accepted by any reasoning organism.

This doesn't yet get us to universes made of mathematics. Tegmark's starting point is that the job of maths is to describe structures. These are not structures we invent – they are inherent in reality. While writing $2 + 2 = 4$ or saying 'if I have two oranges and you give me two oranges I will now have four oranges' are using symbols or words we have invented, the physical reality is what happens to my pile of oranges, described by either of those approaches. There is a tangible mathematical reality in the physical structures involved.

So far, so good. But here's where Tegmark diverts from the mainstream. Working from two premises, that a 'theory of everything' should have no human-specific baggage and that something with a 'complete baggage-free description is precisely a mathematical structure', he makes the leap to 'if you believe in an external reality independent of humans, then you must also believe that our physical reality is a mathematical structure'.

We are very much on a knife edge between science and philosophy here. But if we temporarily accept Tegmark's assertion, this leads us to a particularly obscure but comprehensive multiverse. He suggests that once we have got our heads around a mathematical universe based on a particular set of equations, such as those of the general theory of relativity, we need to ask about all the other equations that don't appear to be represented in our universe. As there is no particular reason for picking the set we have (unless we go back to interference from a deity), then surely, he argues, there must be universes for each possible combination of equations.

Tegmark asserts that the existence of such a mathematical multiverse is non-optional if you accept the idea that mathematics does not just describe reality, but rather *is* reality. As he puts it, the mathematical universe does not exist in space and time, but space and time may exist in it. From this he assumes there must be many other mathematical universes, though most would not support the existence of life.

While I can accept Tegmark's argument that there is something fundamental in physical reality that is mathematical in nature yet more concrete than $2 + 2 = 4$, I am not happy that this is extendable to all mathematics. Numbers, whether cardinal (counting) or ordinal (position in an order), certainly have physical equivalence, as do various other mathematical concepts such as sets, though even they have a distinct logical problem.* However, it doesn't follow that all of reality has such a mathematical basis, nor that all mathematics is directly reflected in physical reality.

A good example of this is that string theory works best if the cosmological constant that describes the expansion or contraction of the universe is negative – unfortunately such a value bears no resemblance to our physical reality, where the constant is (just) positive. Similarly, as we've already mentioned, mathematical dimensions need have no equivalence to physical dimensions.

Many people would also argue that the whole of reality is not reducible to mathematical theory of everything. We know (and Tegmark acknowledges) that there comes a point where reductionism† becomes impractical. So, for instance, a perfect reductionist view of the air in the room you are sitting in would need mathematical descriptions for every molecule's movements, along with all the behavioural oddities of quantum particles, simply to deal with the pressure or temperature. In

---

* A problem with the relationship between set theory and reality is the so-called axiom of choice, a technical assumption that is essential for set theory to work, but that seems to have no reason for working.

† Reductionism refers to the practice of explaining a potentially complex occurrence by being able to describe the behaviour of its most basic units, such as quantum particles and their parameters such as energy and velocity.

practice we can't do that and make do with statistical approximations.

That, admittedly, introduces what Tegmark calls 'human baggage' – it may be impossible to do in practice, but it wouldn't stop the mathematical basis for its existence. Many feel that the experience of sitting in that room, thinking about that air and the associated feelings – perhaps even our consciousness to be able to do that – cannot be dealt with by reductionism. Tegmark's mathematical multiverse, which Brian Greene describes as the 'ultimate multiverse', is not for me.

Falling into a black hole

Meanwhile, back at 'it from bit', we get a different way of looking at things that has the potential to be interesting. For many people, admittedly, it is not strictly 'it from bit' (i.e. there is nothing physical other than the information) but 'bit from it', where the information is the result of physical reality. But it's still an interesting approach, and of importance in assessing the possibility of holographic universes because of the relationship between black holes and information.

One of the fascinating aspects of black holes is imagining what would happen if you were to fall into one. There certainly would be potential problems. Assuming you go feet first, for example, the closer you get to the centre, the stronger the difference would be between the gravitational pull on your feet and on your head. The feet would be pulled more and more strongly compared to the head, resulting in a stretching effect given the entertaining name of spaghettification. But unconnected from this there would also be a point where you passed through the black hole's event horizon.

As we have seen, this is an imaginary sphere around the black hole that is the point at which the gravitational distortion of spacetime would be so great that even light would be unable to escape, as it would be bent back into the black hole. You would not feel the event horizon as you passed through it – and depending on the size of the black hole it may come before or after spaghettification became terminal (or even noticeable). But it would mark a major change as far as the information you contain goes.

You might not consider yourself a repository of information, but this would include both what we usually think of as information – the knowledge encoded in your brain, for example – and the information held in the structure of your body, overall making up your 'bit from it' content. Once you had passed the event horizon, that information would be lost for ever. It wouldn't be seen again. And that's a potential problem, because at the 'bit from it' level, if we lose that information store, we are tampering with the second law of thermodynamics that we already met when considering cyclic universes.

Collapsing in the deepest humiliation

The second law is one of the most fundamental laws of nature. Famously, the English astrophysicist Arthur

Eddington said, 'If someone points out to you that your pet theory of the universe is in disagreement with Maxwell's equations – then so much the worse for Maxwell's equations. If it is found to be contradicted by observation – well these experimentalists do bungle things sometimes. But if your theory is found to be against the second law of thermodynamics I can give you no hope; there is nothing for it but to collapse in the deepest humiliation.'

You could represent the law as 'you can't get something from nothing', but more formally it says that statistically the entropy in a closed system will stay the same or increase. Order won't spontaneously grow without the input of energy. And information and entropy are strongly linked. Entropy reflects the difference between the actual complete state of a system – its total information – and what we know about the system, because there may be many ways to organise its components.

Note, incidentally, that the second law doesn't prevent *useful* information being lost. If a book burns and you just get the resultant ash and gas, the information in that book is no longer usefully accessible. Yet, in theory, the ash and gas still have the capacity to hold as much information as they ever did. The system has not reduced in information content.

Back at our black hole, we know that as the hole absorbs matter its event horizon grows in size – there is a direct relationship between the mass of the hole and the scale of its event horizon. But do its contents truly still have the information capacity that they once did? Or has the disorder totally disappeared into a totally ordered nothingness? Usually we would consider the distinction between being inaccessible and gone to be significant. If I put a book in a safe and throw away the key, the information is inaccessible but it still exists. Yet in the black hole it would be reasonable to say that the information and the capacity to store it has truly gone, because it should never be accessible again.

Black holes aren't black

This paradox would be (possibly) dealt with by Stephen Hawking. Quantum theory tells us that empty space is not truly empty, as over very short periods of time energy levels can fluctuate wildly, providing sufficient energy to generate matter in the form of particle-antiparticle pairs. For example, these might be an electron and its antimatter equivalent, the positively charged positron. Usually, the two particles are almost immediately attracted to each other and annihilate, returning to the fluctuating energy mix. But near a black hole, one particle might get absorbed while the other makes it away.

Although both electrons and positrons have mass and positive energy, it is possible to treat a positron as the absence of a negative energy electron (this is sometimes referred to as a 'hole' in electronics). And for technical reasons the black hole has more chance of 'eating' the negative energy particle. The result is that the positive energy particle emerges as radiation while the negative energy particle reduces the mass of the black hole. Mathematically, this process effectively rescued the second law of thermodynamics, as it ensured that entropy was preserved at the event horizon – and, effectively, with it the apparently lost information.

The result, then, appears to be that the information absorbed by a black hole is not destroyed but rather is encoded in the event horizon, rather as a hologram encodes the information held in the incoming photons of light. In principle this is possible, as the two-dimensional surface of a sphere is capable of holding more information than the three-dimensional volume within it.

If we take the final step, we can envisage that for a finite universe, all the information that corresponds to its quantum particles interacting in it could be represented in the event horizon of a black hole. There is no reason for thinking that this is the case, but it could be. No one has provided an explanation of how that information is stored or accessed – bear in mind that we are only saying that it is there in the same sense that all the information in a book is still 'there' when it's burned to ash and gas – but we have no way of retrieving that information.

There are also some mathematical issues when squeezing an expanding universe like ours, especially if it is infinite in the first place, into a holographic format – but it is just about conceivable. And, of course, if a universe could exist in the event horizon of a black hole then, assuming there is more than one black hole, we have the potential structure for a multiverse.

There is no doubt that it can be fun coming up with all the different kinds of potential multiverse that are mathematically conceivable. But it is also possible to wonder if this is just a kind of highly stylised science fiction. This, then, is the final consideration for our study of the multiverse. Is the concept science at all?

ANGELS ON PINHEADS 11

A caricature of the strange mentality of the medieval worldview is that the philosophers of the day spent their time discussing how many angels could dance on the head of a pin. In reality, although much of what was undertaken did not have the firm foundation in experiment and observation we would expect in modern science, there was a considerable amount of effort put into what we would now consider scientific enquiry, such as the early development of optics in the 'natural philosophy' of the day.

While it's true that thirteenth-century Italian Christian philosopher Thomas Aquinas mentioned the possibility of angels existing in the same place simultaneously (because of their assumed lack of a physical body), this was never a major point of debate. The dancing on a pinhead motif seems to have been dreamed up in the seventeenth century as an attack on earlier unworldly philosophers, before moving into common usage as a metaphor for engaging in pointless debates of philosophical niceties while ignoring practical realities.

Is it science at all?

As we have seen in looking at multiverses, though, there is an aspect of a 'pointless debate' characterisation that has continued into current science, notably in cosmology and the more speculative parts of particle physics and quantum physics. The very driving force behind science is to put forward theories that are evidence-based and can be tested. But some scientists can't resist delving into speculation where there is no evidence – and in some cases where there never will be.

As physicist Sabine Hossenfelder points out in her excellent book *Lost in Math*, it is particularly common in particle physics and cosmology for physicists to dream up whole rafts of theory supported only by mathematics, much of which can never be experimentally confirmed. What *can* be checked is often so expensive to work on that only a very small number of possibilities can ever be examined.

It's the maths that is in the driving seat, which feels wrong. As Hossenfelder points out, string theory, for example, works best if the cosmological constant value that reflects the expansion or contraction of the universe is negative. Unfortunately, it's actually positive, but most string theorists spend their time working with a negative cosmological constant. It makes for more beautiful mathematics – but has nothing to do with our universe. It's also the case that the vast majority of theoretical advances in physics used to be made by individuals, whereas now most theoreticians work in teams. It is tempting to wonder, if a camel is a horse designed by committee, what is a theory developed by a group?

Hossenfelder repeatedly comes back to two measures used to evaluate these untestable theories – beauty, which is inevitably a subjective phenomenon, even though there is some agreement of what is required for beauty – and naturalness, which appears more scientific as it involves numbers, but relies on a bizarre confidence that values in nature that are dimensionless (for example, ratios of masses) should be Goldilocks-like in not being too big or too small, but should have a value of around one.

The physicists that Hossenfelder interviewed for her book (nearly all male), often seem to cling onto these measures without being able to justify them, other than saying that everyone else likes them too. There are some attempts to come up with a defence – one interviewee, for example, suggests that the appeal to beauty is an evolutionary response to a successful theory. But that only seems to demonstrate a weak understanding of the nature of evolution.

Vast amounts of physicist-hours have been put into theories such as string theory, which seems incapable of doing the main job that it is supposed to do, or employed in defending an extension of the standard model of particle physics into a realm known as supersymmetry. This extension of a successful, well-tested theory is required for string theory, yet more and more evidence suggests it is unlikely to be true. If supersymmetry, and the whole zoo of extra particles it predicts, were reflected in reality, we should have seen some of those particles emerging in the Large Hadron Collider at CERN by now – but we haven't.

Hossenfelder demonstrates that clinging to theories past their sell-by date is almost inevitable because physicists are human. If you've spent most of your career on a theory, you don't give it up easily, even though scientists are supposed to love falsification as much as success. And if hundreds of other people are working on a particular theory, they reason, surely it must have some substance? This means that, rather than, for instance, taking the non-discovery of supersymmetric particles as evidence of a need to change direction, there have been arguments for building an even bigger collider, despite the reality that the current machine should have detected them if they existed.

One thing that *Lost in Math* doesn't mention, but may be worth thinking about, is that there are, perhaps, too many theoretical physicists. Hossenfelder points out in a period of about a year when the LHC produced data that looked interesting but turned out to be the result of a statistical fluctuation, 500 papers were published exploring this non-event theoretically, many in top journals. There's nothing wrong with publishing negative results. In fact, scientific publishing all too often fails to cover failures, making the picture given by studies a false representation of reality. But there comes a point when you have to realise that you are backing the wrong horse.

Inevitably, there was a widespread negative reaction from the physics community to *Lost in Math*, but this is not surprising when Hossenfelder was attempting to burst a self-reinforcing social bubble just as powerful as those that surrounds some political views. The knee-jerk reaction is always to deny there's anything wrong – yet here it seems so obviously a case of the emperor's new clothes.

Linking the bubbles

Unfortunately, at least some aspects of multiverse theory fit comfortably into the 'highly speculative, without evidence' bracket and can arguably considered not to be truly scientific. As we have seen, the attempt to use fine-tuning as an argument for the existence of multiple bubble universes or something similar is little more than a misuse of probability. The basic concept of a bubble-like multiverse is that it is impossible to detect one bubble from another. But there are a couple of means that have been attempted to get around the angels-on-a-pinhead nature of this theory. One is by considering a potential behaviour of black holes (a behaviour that is itself highly speculative), and the other is to make use of a potential aspect of quantum computing.

To enable black holes to provide a mechanism to access or communicate with another universe requires a considerable extension of the concept. This requires an Einstein–Rosen bridge, better known (primarily from fiction) as a wormhole. This hypothetical mechanism provides a link between two locations by using a black hole linked to a white hole (a kind of anti-black hole) that would (in principle, as far as the maths is concerned) enable travel between universes. We do need to stress, though, that no one has ever seen a wormhole (or a white hole) and it seems highly unlikely we could either construct a wormhole or find a naturally occurring one leading to another bubble universe. The reality of such a mechanism for establishing the existence of a multiverse feels just as unlikely as a group of angels appearing in your high street to do a pinhead dancing display.

But what of proof by quantum computer?

Proof we live in a multiverse?

Towards the end of 2024 the press was full of the suggestion that Google's new quantum computing chip demonstrated the existence of a multiverse. On 12 December, for example, the *Daily Mirror* newspaper ran with the headline 'Google claims it has discovered staggering PROOF "we live in a multiverse"'. The hype here was not entirely the fault of the newspaper, but was inspired by Google Quantum AI's Hartmut Neven, who was quoted as saying:

> Willow [the chip]'s performance on this benchmark is astonishing: It performed a computation in under five minutes that would take one of today's fastest supercomputers 10^{25} or 10 septillion years. If you want to write

it out, it's 10,000,000,000,000,000,000,000,000 years. This mind-boggling number exceeds known timescales in physics and vastly exceeds the age of the universe. It lends credence to the notion that quantum computation occurs in many parallel universes, in line with the idea that we live in a multiverse, a prediction first made by David Deutsch.

Note, by the way, that Neven does not mention proof. While, as we have seen (page 114), Deutsch did indeed suggest that it might be possible to use some form of quantum AI to detect the existence of 'parallel universes', aka a multiverse, the reality of even his highly speculative theory is more subtle than this. As Deutsch has put it, 'You have to write out many paths for the apparatus – many different histories of it. And if this is applied to different histories of the laboratory as a whole – of the experimenters and so on, then that is the parallel universes interpretation.'

If these 'parallel universes' did exist, it wouldn't seem unreasonable to explain the remarkable abilities of quantum computers, which depend on quantum effects, by thinking that effectively there are multiple processes running in separate universes, combining to provide the super-fast outcome. But the multiple paths in the twin slits or the parallel operations of a quantum computer aren't really the same as the many worlds hypothesis – they just sound a bit similar. Which isn't how science works – it's a mechanism employed by pseudoscience.

The many worlds interpretation, as we have seen, is about each possible outcome of a quantum interaction – for example, each possible location on the screen the twin slit experiment could light up – occurring in a different universe. We occupy a single universe where the particular result observed occurs. But in our universe alone it's still the case that all the possible paths exist as probabilities. To be fair to Deutsch, as we have seen, he envisaged a self-aware quantum computer (hence the appeal to Quantum AI head Neven) that was able to observe its own quantum state. What we know for certain is that Google's Willow chip is not a self-aware computer.

When considering fine-tuning in chapter five it proved helpful to bring in a philosopher to cut through some unscientific thinking by scientific speculators. This demonstrates perhaps the extreme of something that crops up time and again when we have been looking at multiverse concepts. These are, almost by definition, ideas that go beyond what is detectable by experiment or observation. It often feels as if multiverse theory bears some similarities to the invisible dragon theory.

Meet the invisible dragon

In my garage I claim to have an invisible dragon. Can we use science to prove that it does not exist? Clearly it can't be detected by looking for it. It's invisible, after all. But what else could be deployed? Give it a moment's thought. Because of the nature of my invisible dragon, I can pretty much guarantee that whatever you think of, I can render your mechanism invalid.

So, for example, you might scatter the floor of the garage with flour to pick up invisible dragon footprints, or turn the floor into a giant weighbridge to detect its mass. Sorry, my dragon is massless. It won't make impressions or be picked up on weigh scales. Perhaps you might use a temperature probe – but my dragon's physiology means that its surface temperature is always a perfect match to that of its surroundings. How about spraying it with paint or simply prodding the empty space of the garage?* No, my invisible dragon is made up of dark matter particles that do not interact electromagnetically. Ordinary matter will pass straight through it. And so on.

Going ascientific

What this means is that my hypothesis – that there is an invisible dragon in my garage – cannot be falsified. Philosopher of science Karl Popper suggested that for a claim to be scientific it must be capable of being disproven. Scientific theories are usually not absolute truths, but rather represent our best understanding given current data, and should always be susceptible to change. The classic example is the black swan. It was a reasonable theory for European scientists to hold that all swans where white, until someone falsified this theory by discovering a black swan in Australia. But if, like my dragon, it is impossible to ever disprove it, then arguably science has nothing

---

* Prodding dragons is not recommended. Please do not try this at home.

to say on the matter – it's what Sabine Hossenfelder has called ascientific.

I ought to say that Popper's requirement that we should be able to disprove a theory to make it scientific is frequently not easy to fulfil. If an experiment runs counter to theory, the concept is not disproved. There could be problems with the experiment itself. But over time, we can see a growth in evidence supporting the theory, or an increasing belief that it is incorrect. And that evidence cannot be gathered if there is no way to test the theory in the first place.

Without doubt some theories are ascientific – and this isn't meant as an insult. Perhaps the most familiar and most widely held ascientific concept is the existence of one or more deities. Despite the attempts of philosophers and theologians over the centuries, it is not possible to produce a scientific argument for or against the existence of God or gods. This reflects the concept of 'non-overlapping magsteria' devised by the American palaeontologist Stephen Jay Gould. But putting aside theology, we need to ask whether or not multiverses are ascientific too.

This may seem to be a question with an obvious answer of 'yes', but we have to bear in mind that science does develop and what was ascientific in the past can be brought into the fold of science if new evidence comes to light. It certainly would be nothing new for science driven primarily by mathematics to cause consternation among foremost scientists, only to later become widely accepted.

There is a powerful example of this happening in the second half of the nineteenth century. At the time, physics and mathematics were quite distinct disciplines. This may seem strange, given that, for example, Isaac Newton's work was heavily based on geometry and calculus. But Newton and his contemporaries would have considered laws of motion and gravitation as part of the realm of mathematics. Although physics made use of numbers, its findings were purely driven from observation and experiment.

The invasion of pure mathematics

The Victorian Scottish physicist James Clerk Maxwell would take a major step forward in driving physical deductions purely from a starting point of mathematics. In his earlier work, Maxwell, like his contemporaries, relied on analogy with more directly observable mechanical phenomena. So, for example, light would be thought of as acting like a wave. Electricity was thought of as behaving like a fluid. And so on. Analogy allowed very useful theoretical models of reality to be developed. Yet Maxwell wanted to go further. Rather than continuing to make use of models of nature in this way, he would take away the scaffolding of analogy, leaving just the purity of mathematics behind.

Maxwell gives a fascinating picture of how this might be done (amusingly making use of a mechanical analogy to do so):

We may regard this investigation as a mathematical illustration of the scientific principle that in the study of any complex object, we must fix our attention on those elements of it which we are able to observe and to cause to vary, and ignore those which we can neither observe nor cause to vary.

In an ordinary belfry, each bell has one rope which comes down through a hole in the floor to the bell ringers' room. But suppose that each rope, instead of acting on one bell, contributes to the motion of many pieces of machinery, and that the motion of each piece is determined not by the motion of one rope alone, but by that of several, and suppose, further, that all this machinery is silent and utterly unknown to the men on the ropes, who can only see as far as the holes in the floor above them.

Supposing all this, what is the scientific duty of the men below? They have full command of the ropes, but of nothing else. They can give each rope any position and any velocity, and they can estimate its momentum by stopping all the ropes at once, and feeling what sort of tug each rope gives. If they take the trouble to ascertain how much work they have to do in order to drag the ropes down to a given set of positions, and to express this in terms of these positions, they have found the potential energy of the system in terms of the known co-ordinates. If they then find the tug on any one rope arising from a velocity equal to unity communicated to itself or to any other rope, they can express the kinetic energy in terms of the co-ordinates and velocities.

These data are sufficient to determine the motion of every one of the ropes when it and all the others are acted on by any given forces. This is all that the men at the ropes can ever know. If the machinery above has more degrees of freedom than there are ropes, the co-ordinates which

express these degrees of freedom* must be ignored. There
is no help for it.

In Maxwell's example there is no need for a mechanical
model describing what is happening above the ringers'
ceiling, out of sight, to be able to describe how the actions
of the bell ringers translated into sound. Mathematics
alone could be used to model what was happening and
predict outcomes.

Most of the physicists of Maxwell's day were baffled
by his insight. Bear in mind, one of the greatest physicists
of the time (arguably of all time), Michael Faraday, had
hardly any mathematical skill. Faraday wrote to Maxwell
in 1857:

> There is one thing I would be glad to ask you. When a
> mathematician engaged in investigating physical actions
> and results has arrived at his conclusions, may they not be
> expressed in common language as fully, clearly and defi-
> nitely as in mathematical formulae? If so, would it not be
> a great boon to such as I to express them so? – translating
> them out of their hieroglyphics, that we also might work
> on them by experiment.

Faraday considered the mathematical modelling of reality to be on a par with the then near-indecipherable

---

* In physics, 'degrees of freedom' means the number of different
parameters defining the state of a system – if you know all of these,
you can say exactly how it will behave, but if you only know some
of them, you will be limited in your ability to predict its response.

Egyptian hieroglyphics and wanted a translation to be able to connect the theory to experiment. Similarly, William Thomson, later Lord Kelvin, a direct contemporary of Maxwell's and one of the leading physicists of the day, admitted that he never came close to grasping what Maxwell was doing.

The big change that would come with the next generation of physicists was to embrace this mathematical approach, such that, for example, the theoretical physics department at Cambridge is called the Department of Applied Mathematics and Theoretical Physics. Although theory would sometimes be related to and tested against experiment, often theoreticians spend most of their time inhabiting worlds of unconstrained mathematical possibilities where dimensions are innumerable.

Unfettered by physical reality

Nowhere was the move from considering what was known to be real more common than in cosmology, so much so that there was a common jibe in the scientific world that 'there is speculation, and then there's extreme speculation … and then there's cosmology'. Cosmology could never be subject to experiment – at best it was a case of speculating as a result of observation, but often mathematics was allowed free rein.

This is the circumstance we find ourselves in when looking at the different versions of multiverse, whether they be cosmological or interpretations of quantum physics. We have to ask whether it is worth undertaking such work.

It's a difficult question. There is no doubt that unconstrained thinking is sometimes essential to make a breakthrough. As the philosopher of science Thomas Kuhn* pointed out in his groundbreaking book *The Structure of Scientific Revolutions*, science tends to go through periods of gradual small accumulations of knowledge before reaching what he refers to as a paradigm shift, where the old guard are pushed aside and new thinking takes charge.

Leading up to such a shift, existing scientists think that they are just dealing with the details required for extra precision, when suddenly things change in a way that they often find difficult to comprehend. Maxwell's mathematical physics was arguably one such shift, just as quantum physics would be so groundbreaking that major figures behind its development such as Max Planck and Albert Einstein would consider it to have gone too far and could not possibly be correct.

It may be that speculation on multiverses represents the beginning of another paradigm shift (although, if it is, it's taking a long time to get going). And there is nothing wrong with theoreticians playing around with such ideas in the hope that they might at some point find a

---

* We have to take Kuhn's work with a pinch of salt. His idea of paradigm shifts was tied up with a degree of post-modernist strangeness. He did not think that our views changed, but rather the world *itself* changed, as there was no fixed reality, just what we saw it to be. So, for instance, he considered Newton's 'gravity' to be a totally different thing to what we mean by the word. But you can put aside the dippier aspects of the thinking of the period and still see the usefulness in the idea of sudden changes of viewpoint.

way to anchor them in reality and allow the theories to be tested by experiment and observation. But until that time, scientists and science writers need to be careful not to overhype their ideas. It's something that all too frequently occurs.

The danger of overhyping

I have previously pointed out a delightful (indeed, hilarious) set of news stories from 2013 that illustrate this overhyping to perfection. We heard from Fox News: 'Star Wars lightsabers finally invented'. (I love the use of 'finally' here.) The generally more sober *Guardian* newspaper gave us 'Scientists Finally Invent Real, Working Lightsabers', while *Time* magazine told us that 'MIT, Harvard scientists accidentally create real-life lightsaber'.

These stories were based on a press release in which Professor Mikhail Lukin of Harvard University said, 'It's not an inapt analogy to compare this [discovery] to light sabers.' In fact, that was exactly what it was. The team had managed to make two photons of light, which usually ignore each other, briefly interact in an ultra-cold quantum substance known as a Bose–Einstein condensate. That's it. No lightsabers.

When scientists, university press offices and the press conspire in this way to overhype scientific discoveries it risks devaluing the trust the public has in their announcements. The same is true of the barrage of 'scientists show' or 'new study shows' news pieces we get bombarded with every day. Such stories often describe an inconclusive study that merely indicates it's worth putting more effort in to find out more about a topic, but they are treated as if they have made a proven discovery.

Multiverses haven't even reached that stage – studies aren't showing anything yet. It's just that there is theory that, in principle, might make future studies possible. Even a multiverse enthusiast like Brian Greene admits that multiverses may never be testable in the normal sense. He commented, 'No doubt you've noted that even the most sanguine projections suggest that predictions emerging from a multiverse framework will have a different character from those we traditionally expect from physics.' To accept a physical theory, we usually look for a high level of accuracy. If the measurement is statistical, as many now are, physicists usually expect a '5 sigma' result.

What is significant in science?

This statistical approach was required, for example, to accept the discovery of the Higgs boson. No one saw a Higgs boson, or any direct evidence of one. Instead, its existence was statistically implied by the data produced by the Large Hadron Collider. Anyone who looks in some detail at scientific results may have come across either p-values or sigmas being used to determine the significance of an outcome – but in considering how convincing a result is it's important to understand what these values represent, and why there is a huge disparity between practice in the social sciences and physics.

These are statistical measures that determine the probability of the results being obtained if the 'null hypothesis' were true – which is to say if the effect being investigated does not exist. The social sciences, notably psychology, usually consider the marker for statistical significance to be a p-value of less than 0.05, while in physics the aim is often to have a 5-sigma result.

Both these measures depend on creating a probability distribution, showing the likelihood of different values occurring. The p-value is a direct measure of the probability of getting the reported results if the null-hypothesis applies. So, a p-value of 0.05 means there is one in twenty (1/20 = 0.05) chance of this happening. Sigmas effectively measure the same thing, but in terms of a measure called standard deviation that shows how spread out the distribution is.

It might seem odd not to use the more straightforward p-value, but the reason that sigmas tend to be used in physics is that the p-value equivalent becomes very small at the levels that physicists use. CERN, for example, does work with p-values, but converts them to sigmas for easier communication. Here's a look at equivalent values:

Sigma	P-value	Cliff measure
2	0.05	Whiff
3	0.003	Evidence
4	0.0001	Annoying
5	0.0000003	Discovery

The 'Cliff measure' used above is a humorous interpretation of sigmas given by English particle physicist Harry Cliff,* in his book *Space Oddities*. (Here, 4 sigma is called 'annoying' because it's almost a discovery. Most social scientists would consider it an incredibly good result.) Arguably this is a more effective description of the value of different levels than the way statistical significance is usually regarded in the social sciences.

Choosing 2 sigma/p-value of 0.05 to indicate being statistically significant was an arbitrary decision, plucked out of the air by English mathematician Ronald Fisher in 1925. However, it should be seen as nothing more than a note that something is worthy of proper investigation – Cliff's whiff – rather than an indicator that the outcome is accepted science. One reason the newspapers are always carrying details of studies showing something apparently remarkable that turns out to be highly unlikely is that they interpret a whiff as a discovery.

Such has the focus been on getting a p-value below 0.05, there has in the past been a significant amount of 'p-hacking' – manipulating the data produced in a study with the intention to get the result below this critical

---

* When mentioning Harry Cliff I need to add an apology to him. Some years ago, I wrote an article about an attempt to turn the Cavendish Laboratory in Cambridge into a worthy memorial of the work that was undertaken there. I interviewed Harry Cliff for this, but accidentally referred to him as Harry Webb in the published article. Fans of early British pop music may suspect the accident was due to being aware that the singer Cliff Richard was born Harry Webb. Dr Cliff found this amusing.

level. The simplest way to do this is to slice up the data twenty different ways – at least one is likely be significant purely by chance. But Fisher certainly never intended 0.05 to be an indicator of a real discovery. Remember, a p-value of 0.05 means that there is a 1 in 20 chance of getting these results when the effect doesn't exist. It's a better probability than Russian roulette (p-value 0.17), but it's still hardly something you would want to risk your life on.

Why, then, is there such a disparity between the social sciences and physics? Mostly because it isn't practical to have sufficient experimental subjects or experimental runs to come close to a 5-sigma outcome in, say, psychology. As a result, the social sciences can't hope for equivalent degrees of apparent certainty. However, there is a strong feeling that the social sciences could do better – perhaps aiming for 3-sigma before they get excited. And it does mean that the outcomes of social sciences studies should arguably carry a health warning and be better reported in terms of the risk of their misattributing an outcome to a particular cause.

One final consideration – even 5-sigma results can be wrong. Scientists can make a mistake with the maths. And there could be confounding factors too – a great example is the BICEP2 study apparently showing the impact of inflation, which we met in chapter seven. BICEP2's abandoned discovery was at the 5.9 sigma level. Yet it proved to be a misinterpretation rather than a discovery. There is always the possibility that scientists have not allowed for a factor that has distorted the results – something that sadly tends to disappear from popular science/news reporting where outcomes are often stated is if they are fact.

Probability and statistics can be hard to get our heads around – but when scientific results are reported, it is essential that this particular aspect should be carefully explained up front. To have confidence in scientific results, we need to know what the limitations of a particular study are. Yet, returning to Brian Greene's admission that results would have to be accepted of 'a different character', he notes, 'From where we now stand,* it seems that multiverse predictions will never reach this standard of precision.'

'It's vaguely possible that something is true' doesn't make for the kind of exciting headlines scientists (or the media) like to see. It's fine to carry on theorising and speculating. After all, the development of theories is cheap. But until we have experimental or observational verification, we should not be shouting about multiverses from the rooftops. And before we spend vast sums on upgraded colliders searching for supersymmetric particles or next generation observatories looking for cosmic microwave variations, we need better experimental data on the small scale to justify these expenditures.

Can we move beyond clickbait?

If you filter out the vast majority of news stories with 'multiverse' in the title (mostly those relating to the Marvel Cinematic Universe), you end up with a whole

---

* Brian Greene's comment 'From where we now stand' dates back to 2011, but if anything we are even less confident now.

list of sensationalist headlines. Here's a few I came across in a quick search:

- Aliens from a Parallel Universe May Be All Around Us – And We Don't Even Know It, Study Suggests (*Popular Mechanics*) – *They are probably playing with my invisible dragon. By definition **anything** can be happening in a parallel universe we can't detect, but what does it tell us?*
- Could we travel to parallel universes? (*Live Science*) – *No.*
- Why scientists think the Multiverse isn't just fiction (*Big Think*) – *Many don't think this. And the sensible ones who like multiverses only think it **might** not be fiction. This story was based on eternal inflation.*
- Why do people think NASA has discovered a 'parallel universe?' (*Newsweek*) – *They don't. This was reporting an unfounded claim that an initially puzzling neutrino behaviour could be caused by interaction with a parallel universe where time ran backwards. There were plenty of less bizarre explanations and the claim was never taken seriously.*
- Our reality seems compatible with a quantum multiverse (*New Scientist*) – *It's also compatible with my invisible dragon. It doesn't mean that it exists.*
- We are closer than ever to finally proving the multiverse exists (*New Scientist*) – *No, we aren't. The strange thing is that (like many such headlines) there is nothing in the article to suggest we are closer than ever to having proof of something that almost certainly can't be proven.*

These stories aren't going to go away – they are the clickbait of science journalism. But it doesn't mean we have to abandon speculating about multiverses. Just that if we want to take an honest, balanced look at the subject, we would benefit from more thoughtful reporting.

The best we can say as yet is that there is no good evidence that multiverses exist, but we will certainly go on speculating and searching for that evidence. And for many of us, this is just as much fun as making a real discovery.

FURTHER READING

Big Bang

Before the Big Bang, Brian Clegg (St Martin's Press, 2009) – wide ranging look at the origins of the universe, including brane universes.

Entanglement

The God Effect, Brian Clegg (St Martin's Press, 2006) – an exploration of the nature of quantum entanglement and how it has been examined theoretically and experimentally.

Fine-tuning

A Brief History of Time, Stephen Hawking (Bantam Books, 1998) – more detail on Hawking's views on fine-tuning and the possibility of the universe.

Why?, Philip Goff (OUP, 2023) – philosopher Goff shows why the probabilistic argument for the universe doesn't work in a book suggesting the universe has a purpose.

Gravitational Waves

Gravitational Waves, Brian Clegg (Icon Books, 2018) – an introduction to the long search for gravitational waves and how our newfound ability to detect them opens up a whole new way to look at the universe.

Gravity

Gravity, Brian Clegg (St Martin's Press, 2012) – an exploration of the nature of gravity, why it's so weak a force, attempts to quantise it, and the general theory of relativity.

Infinity

A Brief History of Infinity, Brian Clegg (Constable & Robinson, 2003) – explore the mind-bending nature of infinity in far more depth.

Mathematical Universes

Our Mathematical Universe, Max Tegmark (Penguin Books, 2015) – gives a very readable if sometimes baffling description of the different possible options for multiverses and why the ultimate multiverse is one based on different mathematical structures.

Maxwell and Mathematical Physics

Professor Maxwell's Duplicitous Demon, Brian Clegg (Icon Books, 2019) – a scientific biography of James Clerk Maxwell, including the resistance of physicists to his mathematical approach.

Olbers' Paradox

The 1987 study showing 'many workers appear to be still ignorant' of the solution to Olbers' paradox was *The Extragalactic Background Light and a Definite Resolution of Olber's Paradox* by P. Wesson, K. Valle and R. Stabell in *The Astrophysical Journal*: 317, 601–606.

Parallel Universes

If you would like to dig into multiverse possibilities at a deeper level (though with less scepticism about speculation), Brian

Greene's *The Hidden Reality* (Pelican, 2011) gives a good deep dive, as seen by an enthusiast. It is a little dated now, but gives a useful picture of how scientific theoretical speculation can be justified.

Physical Constants

In *Just Six Numbers* (Weidenfeld & Nicolson, 1999), Martin Rees gives an insightful view into the nature and scale of six of the most important physical constants.

Probability

Dice World, Brian Clegg (Icon Books, 2013) gives a good introduction to the fascinating oddities of probability and randomness.

Quantum Interpretations

Lee Smolin's excellent book *Einstein's Unfinished Revolution* (Allen Lane, 2019) takes the reader through various interpretations of quantum theory, though it is firmly in favour of 'realist' interpretations and shreds the many worlds hypothesis.

Roger Bacon

Roger Bacon, Brian Clegg (Constable, 2013) – the remarkable life and work of this thirteenth-century proto-scientist.

Sets

A Brief History of Infinity, Brian Clegg (Constable & Robinson, 2003) – set theory is fundamental to our understanding of mathematical infinity and is covered in some detail here.

Speculative Science

Lost in Math, Sabine Hossenfelder (Basic Books, 2020) is an effective counter to the tendency of a minority of scientists to indulge in speculation that has no means of being tested.

INDEX